THE UNIVERSE AND THE TEACUP

THE UNIVERSE AND THE TEACUP

The Mathematics of Truth and Beauty

K. C. COLE

HARCOURT BRACE & COMPANY
New York San Diego London

Library of Congress Cataloging-in-Publication Data
Cole, K. C.
The universe and the teacup:
The mathematics of truth and beauty/K.C. Cole.—1st ed.
p. cm.
Includes bibliographical references
and index.
ISBN 0-15-100323-8
1. Mathematics. 2. Truth (Aesthetics)
3. Aesthetics. I. Title
QA36.C65 1998
510—dc21 97-22338

Text set in Minion
Designed by Kaelin Chappell
First edition
E F D

For Frank

Contents

Acknowledgments

Ultimately, the credit for this book has to go to my editor, Jane Isay, who first told me it was a math book. Initially, that came as a complete surprise. I'd come in to talk with her about a bunch of ideas I'd been tossing around for years, inspired by a great number of friends, colleagues, authors of books and essays whom I'd never met, and scientists I'd talked with only on the phone. Yet somehow, I didn't have the perspective to see that the one common thread tying it all together was mathematics. Thank you, Jane.

The other influences encompass almost everyone who ever had an impact on me, so I won't even try to be inclusive. However, a great number can be found listed in the bibliography at the end of this book; a great many others are the scientists and mathematicians who are represented in the text only by an odd quote here and there but whose advice and ideas permeate the whole. Thank you, all, for all those hours of conversations, questions, requests. A special thank you also to Cathleen Morawetz, who carefully went through an early outline of this book (before I knew it was about the mathematics of truth) and offered caveats and encouragement I took very much to heart.

I am most grateful to those who read the entire manuscript for me, in various drafts, offering advice, insights, additions and sometimes simply saving me from stupid mistakes. Thank you: Virginia Barber, Keith Devlin, David Goodstein, Haim Harrari, Roald Hoffmann, Gerald Holton, Tom Humphrey, Patty O'Toole. Thank you, also, to those who rendered the same indispensable service on portions of the book: Susan Chace, Elsa Feher, Adam Frank, Liz Janssen, and Lawrence Krauss. Obviously, whatever stupid mistakes remain are mine.

My editor at the *Los Angeles Times,* Joel Greenberg, helped hone and challenge my ideas as subjects I was exploring for the book inevitably seeped into articles I then wrote for the paper (and eventually, vice versa). Thank you, Joel.

I couldn't have written the book without the continual support of my friends—especially Mary Kay Blakely, Susan Chace, Evelyn Renold, Patty O'Toole, Claudette Sutherland, and Mary Lou Weisman. Thanks to my children, Pete and Liz Janssen, for enriching my life in myriad ways and making me wiser. And a huge thank you to my father, Bob Cole, for always being there.

I'd also like to mention the indirect but important inspiration I received from two writers' organizations I'm proud to be part of: PEN Center West, for reminding me that writers throughout the world pay dearly (sometimes with their lives) for a career I take for granted, and the women of JAWS, for high standards and courage.

Finally, thanks to my agent, Ginger Barber, not only for her thorough and useful commentary but also for setting up that conversation with Jane Isay.

Introduction

THE SENTIMENTAL FRUITS

At the fundamental level nature,
for whatever reason, prefers beauty.

—physicist David Gross, director of the Institute for Theoretical Physics at University of California, Santa Barbara

Mathematics seems to have astonishing power to tell us how things work, why things are the way they are, and what the universe would tell us if we could only learn to listen. This comes as a surprise from a branch of human activity that is supposed to be abstract, objective, and devoid of sentiment.

Yet, the way we view ourselves is closely connected to what we know (or think we know) about objective aspects of nature. Math tells us truths not only about how gravity works (the better to build bridges), but also universal truths that influence how we think and feel (the better to build societies). Physicist Frank Oppenheimer liked to call these the "sentimental" fruits of science.

True, math does all the things we learned in school: building bridges and balancing checkbooks and calculating the odds of winning the lottery. But it also sheds light on those muddles of the mind that keep not only scientists up at night, but also artists and actors and poets and schoolteachers and psychologists and lovers and parents: How can we make sense of nature, including human nature? What is the nature of truth?

People search for the answers in God and in equations (sometimes both at the same time), by writing plays and studying ants. Curiously, the same methods of thinking that helped reveal light as an undulating electromagnetic field can also help sort out the causes of various social problems. The same approaches to proof that physicists used to establish the reality of a particle called the top quark are brought to the courtroom in the trial of O.J. Simpson.

It's a heady notion: Mathematics—that seemingly dry stuff—has so much relevance to the deep philosophical ideas that are the foundations of society. By learning how it works, we can get a better grip on everything from obscure aspects of physics to methods of fashioning fairer divorce settlements.

That's one reason I've tried in this book to directly relate ideas from mathematics to problem solving in unlikely places—from life on Mars to the riddle of the Unabomber. It's an attempt to demonstrate how mathematics informs the kinds of questions people really think and worry about. If I could accomplish one thing in this book, it would be to show that an interest in the quality of life is in no way diminished by quantitative arguments. Quantity and quality are inseparable. Scientists and mathematicians, as well as saints and philosophers, search for the fundamental how's and why's of existence. And although they have different standards of evidence and proof, quantitative insights do help us understand qualitative problems.

Of course, mathematical tools do not substitute for the insights of artists and actors and economists and psychologists and historians and writers and spiritual leaders. But they can supply badly needed fresh perspectives.

This book is structured in five (not equal) parts. In the introductory chapter (What's Math Got to Do with It?), I present the idea that mathematics is not about numbers so much as it is a way of thinking, a way of framing questions that allows us to turn things inside out and upside down to get a better sense of their true nature. Mathematicians know this, of course, but most people outside the profession do not. The chapter tours some of the unexpected territory covered by mathematics—from daily headlines to the Golden Rule.

The first section (Where Mind Meets Math) demonstrates some of the reasons we need math to help sift through the confusion. In the first place, numbers don't speak for themselves, because our all too human brains get in the way. Certain kinds of relationships that ought to be plain to everyone simply can't penetrate the veil that physiology and experience puts between knowledge and truth. Indeed, these mental filters make it difficult (perhaps impossible) for human brains to perceive things the way they really are (whatever that is). They are necessary parts of human psychology and physiology, so there is no point trying to "cure" them. However, it helps a great deal to be aware of them—in the same way as it helps if your car's steering pulls to the left, to compensate by pulling to the right.

The second section (Interpreting the Physical World) explores some of the obstacles to clear seeing that are thrown up by aspects of physical reality itself (not that the muddiness in our minds can ever be completely separated from the messiness of reality). Signals scrambled by persistent interference and changing context, qualities that melt into quantities before our eyes (and vice versa), complex webs of influences that can be impossible to untangle, the elusiveness of observation, and the hazards of prediction—all make the art of getting sense from information a challenge even for the most mathematically adept.

The third section (Interpreting the Social World) gives a taste of how math has illuminated human questions such as fairness. For example, a branch of mathematics called game theory suggests that following the Golden Rule is not only a moral way to behave, but also an effective strategy for getting results.

The fourth and longest section (The Mathematics of Truth) is the heart of the book's premise. It's about some of the ways that mathematics can (and does) frequently reveal surprising fundamental relationships—between causes and effects, for example, evidence and proof, truth and beauty. The juiciest part of all—the payoff (at least for this writer)—is the story of how a young mathematician named Emmy Noether figured out how to make Albert Einstein's general relativity consistent by showing the link between symmetry and the fundamental, unchanging laws of nature. In other words, the same properties that make a snowflake appealing underlie the laws that control the universe. Truth and beauty are two sides of a coin.

Chapter 1

WHAT'S MATH GOT TO DO WITH IT?

Understanding is a lot like sex. It's got a practical purpose, but that's not why people do it normally.

—*Frank Oppenheimer*

Finding out what's true is a central passion of human activity. It's a question that dominates the stage and the dinner table, the classroom and the courtroom, the scientific laboratory and the spiritual retreat. And yet, with the explosion of information reverberating in our brains, it becomes harder and harder to hear the clear ring of truth through the competing facts and philosophies.

As it turns out, mathematics offers a singular set of tools for seeing truth. Indeed, it brings surprising clarity to an astonishing range of issues, from cosmic questions (the fate of the universe) to social controversy (O.J.'s guilt) to specific matters of public policy (race and IQ scores).

People outside the sciences rarely pick up these tools—in part because math seems intimidating. Even if people are aware that such tools exist, they don't know how to apply the tools to things they care about.

But mathematics already underlies many of society's most-cherished political and social inventions: Ideas about cause and effect, fairness and justice, selfishness and cooperation, balancing risks, spending on welfare or national defense, even the nature of scientific discovery itself.

True, our ideas about the physical and social world do spring from sources other than numbers: religion, history, family, psychology. We accept the "truths" revealed by these sources as intuitively commonsensical, or obviously right; our Declaration of Independence describes them as "self-evident."

But math—that most logical of sciences—shows us that the truth can be highly counterintuitive and that sense is hardly common.

Mathematics is a way of thinking that can help make muddy relationships clear. It is a language that allows us to translate the complexity of the world into manageable patterns. In a sense, it works like turning off the houselights in a theater the better to see a movie. Certainly, something is lost when the lights go down; you can no longer see the faces of those around you or the inlaid patterns on the ceiling. But you gain a far better view of the subject at hand.

William Thurston, the director of the Mathematical Sciences Research Institute (and by some accounts the world's greatest living geometer) calls math a kind of "mindware." It allows us to see and articulate concepts we can't handle in any other way. Ingrid Daubechies—the MacArthur Award–winning Princeton mathematician who resurrected wavelet analysis (a tool for doing everything from storing fingerprints to seeing stars)—says it's akin to poetry: a way of taking a big idea and condensing and honing it until it communicates exactly the right information.

Mathematics can function as a telescope, a microscope, a sieve for sorting out the signal from the noise, a template for pattern perception, a way of seeking and validating truth. It is a lens that can clarify the obscure, or obscure and distort what was seemingly clear. It can take you into the core of a star or to the edge of the universe, give you the outcome of an election or the result of pumping carbon dioxide into the atmosphere for a hundred years. You can extrapolate to the end of time, or back to its beginning. You can get there from here.

Mathematicians do not see their art as a way of simply calculating or ordering reality. They understand that math articulates, manipulates, and discovers reality. In that sense, it's both a language and a literature; a box of tools and the edifices constructed from them.

Once I was flying in a plane back from the Boston area, where I had been talking with a cosmologist at MIT about the universe and all that. I looked down from my window and saw islands that were clearly connected under the shallow water by strips of land. On the ground, those links would have been invisible, the islands completely unconnected. From the air, the paths between them were laid out as clearly as road maps. There's a reason, I thought, that a lot of fundamental physics requires looking in higher dimensions. You can see more from an elevated point of view.

In the same way, the tools of mathematics allow one to see otherwise invisible patterns and connections. Mathematics has revealed hidden trends (HIV infection), new kinds of matter (quarks, dark matter, antimatter), and crucial correlations (between smoking and lung cancer). It does this by exposing the bare bones of a situation, overcoming the commonsense notions that so often lead us astray. Math allows you to strip off the coverings and get right down to the skeleton. What is going on underneath that accounts for what you see on the surface? What's holding it up? If you dig deep enough, what do you find?

In some sense, the unfolding story of the universe is a history of finding hidden connections. The nature of light was discovered when a certain number (the speed of light) kept popping out of equations linking electricity to magnetism. Light was exposed as an electromagnetic fluctuation—an understanding that allowed experimenters to go looking for others of its same species. Radio signals, for example, ride on light that vibrates more slowly than the eye can see; X rays vibrate faster.

Equations speak volumes, teasing out economic trends, patterns of disease, growth of populations, and the effects of prejudice and discrimination. Math produces a quite literal expansion of consciousness. It allows us to see more. With these tools, we can extrapolate into the future (but there are hazards) and see invisible things (curved space).

"What do we really observe?" asked Sir Arthur Eddington in 1959, summing up the lessons of the century's recent revolution in physics: "Relativity theory has returned the answer—we only observe relations. Quantum theory returns another answer—we only observe probabilities."*

What we observe, in other words, are mathematical relationships.

Since mathematics is so good at exposing the truth, it's curious how often it's used to perpetuate misunderstandings and lies. Math has power because we give more weight to numbers than we do to words. "Figures often mislead people," says mathematician Keith Devlin. "There is no shame in that: words can mislead as well. The problem with numbers is our tendency to treat them with some degree of awe, as if they are somehow more reliable than words. . . . This belief is wholly misplaced."

People often look to mathematics as an objective line of argument that will rescue them from the uneasiness of ambiguity. If only we put things in terms of numbers, we hope, perhaps truth will out. But math only articulates these ambiguities; it is no lifeboat out of

*Quoted in Richard Gregory's *Mind in Science.*

the sea of confusion—only the buoy that marks the shoals. After all, it was a mathematical theorem (Gödel's theorem*) that proved some truths can't be reached by the road of pure logic at all.

A prime case of intimidation by the numbers is the book *The Bell Curve*, a treatise so controversial that a half dozen books were published in response. Written by Charles Murray of the American Enterprise Institute and the late Richard Herrnstein of Harvard, the book wheels out an arsenal of mathematical artillery to bolster the proposition that intelligence is mostly inherited, that blacks have less of it, and that little can be done about it. Reviewers—not to mention readers—admitted to shell shock in the face of such a barrage of statistics, graphs, and multiple-regression analyses.

Yet the fearless few who plunged into the statistics headlong found that the numbers which seemed to speak so clearly swept crucial qualifications under the rug, making much of the mathematics meaningless.†

The question I get asked most frequently is: How can you ever find out what's true short of becoming a mathematician yourself? The answer is: You don't have to. You merely need the confidence to ask the questions that were probably on your mind anyway. Such as: How do you know? Based on what evidence? Compared to what else? Like the woman who spent a day exploring exhibits at the Exploratorium in San Francisco—then went home and wired a lamp. There was nothing in the world-renowned science museum that taught her how to wire a lamp. What she found there was simply the belief in her own abilities to figure things out.

Used correctly, math can expose the glitches in our perceptual apparatus that lead to common illusions—such as our inability to perceive the true difference between millions and billions—and give us relatively simple ways of protecting ourselves from our own

*See The Burden of Proof.

†See Chapter 12, "The Truth about Why Things Happen."

ignorance. As the physicist Richard Feynman once said: "Science is a long history of learning how not to fool ourselves." A knowledge of the mathematics behind our ideas can help us to fool ourselves a little less often, with less drastic consequences.

In short, math matters—a lot more than most people think. We have to make life-and-death decisions based on what numbers tell us. We cannot afford to remain dumb about mathematical ideas simply because we hated them in high school—any more than we can remain dumb about computers, or AIDS. Mathematics is essential, not peripheral, knowledge.

As someone who started out interested in social questions, I am particularly impressed at the power of math to help sift through evidence and decide what is true in a wide variety of situations. Some of the tools may be obvious (like probability) while others are more subtle and even obscure (like the relationship between symmetry, truth, and things that never change, no matter what).

Many different kinds of truths lie in numbers, and exploring them is the purpose of this book. What does it mean when one number can be correlated with another? Say: IQ and intelligence, or math scores with big feet? If one thing makes another thing more probable, is it fair to call it a cause? What is the most effective strategy for winning at games? Is endless economic growth really a good thing (or even possible)? Was there life on ancient Mars? What's the fairest way to divide the national budget, or the best way to survive a game of "chicken"? What is the probability of getting killed by a terrorist? Getting married after forty? Running into your brother-in-law in Manhattan? In Nome? What, if anything, do these numbers we attach to things mean?

No doubt about it, mathematics embodies great power. It's no wonder that the physicist Sir James Jeans concluded: "The Great Architect of the Universe now begins to appear as a pure mathematician."

At the same time, it is far from foolproof. Like all science, it grows and thrives in cultures and is heavily influenced by their peculiarities. This book focuses on various mathematical guides to the truth that can be applied to a wide range of questions, from issues in the news to matters of purely philosophical or aesthetic interest.

What I personally like best is the way that truth and beauty come together in the work of Emmy Noether and Albert Einstein: How deep truths can be defined as invariants—things that do not change no matter what; how invariants are defined by symmetries, which in turn define which properties of nature are conserved, no matter what. These are the selfsame symmetries that appeal to the senses in art and music and natural forms like snowflakes and galaxies. The fundamental truths are based on symmetry, and there's a deep kind of beauty in that.

The journey begins here.

PART I

Where Mind Meets Math

Mathematics did not appear out of nothing and nowhere—as some cosmologists believe the universe did. It was created (or discovered, if you prefer) by human beings. As such, math reflects many aspects of humanity, including physical characteristics, psychology, and culture. The ways our brains and bodies work have molded not only the study of mathematics, but also our everyday perceptions of quantitative things. It is, after all, human nature—in its broadest psychological and physiological sense—that creates the sophisticated mathematics that revealed curved space-time and quarks and that led to the creation of everything from computers to gene therapy. This same human nature, however, limits our ability to understand phenomena that may be critical to our survival as a species—including risks, population growth, and national budgets.

In what has become a recurring theme in human evolution, the same strategies that serve as stepping-stones to some truths become the obstacles we trip over in pursuit of others in different contexts.

To take an obvious example of how mathematics is shaped by human physiology, most number systems used in societies around the world are built on multiples of ten because virtually all human beings are born with ten fingers and ten toes. (Some cultures expand upon this idea by making use of wrists, elbows, shoulders, and chest.)

Less obvious, our brains appear to be calibrated rather like the scales used to measure the strength of earthquakes, where a small increase on the scale (say, from a 7 to an 8 magnitude) designates an enormous increase in destructive power (on the order of ten times stronger). This may well account for the inability of people—no matter how well schooled in mathematics—to comprehend the true difference between a million and a billion.

In addition, the world beyond our physical bodies is sculpted by forces that produce omnipresent mathematical objects. Geometry really does grow on trees. Because of the way gravitational, electrical, and nuclear forces operate, everything Moon size or larger is round or roundish. Water always falls from a fountain in parabolas. Soap bubbles meet at 120-degree angles, and the two hydrogens and oxygen in water molecules meet at 105 degrees—giving shape to bubbles and snowflakes. Trees and blood vessels and rivers all branch in strikingly similar ways.

By the same token, our measures and notions of time are based on the revolutions of our planet and the time it takes to wind its way around the Sun. We orient ourselves in space along Earth's spin axis (north-south) and magnetic poles. We think of down as a fixed direction, although down for me is up for someone living on the other side of the globe. In truth, down is the direction of the greatest pull of gravity. If you are alone in space, there is no such thing as down, up, east, or west.

Many mathematical operations—like adding and subtracting—directly derive from our physical experience: adding two apples or cutting a pie into thirds or figuring out the circumference of a circle from its diameter.

But mathematical concepts also go beyond experience. The perfect circles and right angles favored by geometers do not exist in the natural world. Numbers can do things that things cannot. There is an old tale that illustrates how some things just don't add up: A man stands on a street corner begging for change by holding up a sign that reads: 2 wars; 1 leg; 2 wives; 3 children; 2 wounds. Total: 10. And things can do things numbers cannot. If you add hydrogen to oxygen in a 2 to 1 proportion under the right conditions, you do not get three units of gas; you get water.

Still, almost every attempt to take numbers beyond experience has met with extreme cultural backlash. When negative numbers were first introduced, people thought they were absurd. Since it made no sense to have minus two apples, what could minus two possibly mean? The introduction of zero was as fiercely contested as the Copernican Sun-centered view of the solar system. And if the stories about the Pythagoreans are true, people actually lost their lives as a result of the discovery of irrational numbers—like pi—that cannot be expressed as fractions. Indeed, our use of the word irrational*—meaning completely nonsensical and off-the-wall—reflects quite accurately how people felt about these numbers when they were first discovered.*

Mathematicians today rely on all kinds of strange objects that were once thought completely out of the bounds of common sense: various kinds of infinities; imaginary and transcendental numbers; higher dimensional geometries; and so forth.

But however far away mathematics gets from human experience, our physical world—including our own physical makeup—continues to play a pivotal role in how we perceive mathematical ideas.

Chapter 2

EXPONENTIAL AMPLIFICATION

The greatest shortcoming of the human race is
our inability to understand the exponential function.

—*physicist Albert A. Bartlett*

Consider the extreme difficulty we have with very large or very small numbers. Anyone who has ever mixed up a billion and a trillion knows that after a while, all big numbers begin to look alike. Daily, we are bombarded with incomprehensible sums:

The national debt has grown to trillions of dollars. The Milky Way galaxy contains 200 billion stars, and there are 200 billion other galaxies in the universe. The chemical reactions that power everything from fire to human thought take place in femtoseconds (quadrillionths of a second). Life has evolved over a period of roughly 4 billion years.

What are we to make of such numbers? The unsettling answer is, not much. Our brains, it appears, may not be engineered to cope with extremely large or small numbers. Douglas Hofstadter coined the term "number numbness" to describe this syndrome, and almost everyone suffers from it. After all, it's so easy to confuse a million and a billion; there's only one lousy letter difference. Except that a million is an almost imperceptible one-thousandth of a billion—a teeny tiny slice.

No one, apparently, is immune. As Donald Goldsmith pointed out in the *Wall Street Journal,* President Bill Clinton managed to lose track of 90,000 physician visits in a speech on health care. He multiplied 500 children by 200 doctors and came up with 10,000 visits, 90,000 short.

All of us have trouble grasping how inflation at 5 percent can cut income in half in a decade or so, or how a population that's growing at even 2 percent can rapidly overtake every inch of space on Earth. From the incredible shrinking dollar to the explosive power of nuclear bombs, things add up in ways that humans find hard to get a handle on. And yet, the consequences of this built-in number blindness are enormous.

If we can't readily grasp the real difference between a thousand, a million, a billion, a trillion, how can we rationally discuss budget priorities? We can't understand how tiny changes in survival rates can lead to extinction of species, how AIDS spread so quickly, or how small changes in interest rates can make prices soar. We can't understand the smallness of subatomic particles or the vastness of interstellar space. We haven't a clue how to judge increases in population, firepower of weapons, energy consumption.

Fortunately, scientists and mathematicians have come up with all manner of metaphors and tricks designed to give us a glimpse at those huge and tiny universes whose magnitudes seem quite beyond our comprehension. University of California, Berkeley, geologist Raymond Jeanloz, for example, likes to impress his students with the power of large numbers by drawing a line designating zero on one end of the blackboard and another marking a trillion on the far side. Then he asks a volunteer to draw a line where a billion would fall. Most people put it about a third of the way between zero and a trillion, he says. Actually, it falls very near the chalk line that marks the zero.

Compared to a trillion, a billion is peanuts. The same goes for the difference between a millionth and a billionth. If the width of this page represents a millionth of something, then a billionth of it would be much less than a pencil line.

S. George Djogvski, writing in Caltech's *Engineering and Science,* offers this analogy to help us imagine the vast distances of space. If the Sun were an inch across and five feet away from our vantage point on Earth, "the solar system would be about a fifth of a mile across. The nearest star would be 260 miles away, almost all the way to San Francisco [from Los Angeles], and our galaxy would be 6 million miles across. The next nearest galaxy would be 40 million miles away. At this point, you begin to lose scale, even with this model—the nearest cluster would be 4 billion miles away, and the size of the observable universe would be a trillion miles. If you were to ride across it at five dollars per mile, you could pay off the national debt."

Not that we can comprehend national debt any better than these numbers. The late physicist Sir James Jeans—a great popularizer of Einstein's theories—wrote about how seemingly impossible it was for people to imagine a range of sizes that goes "from electrons of a fraction of a millionth of a millionth of an inch in diameter, to nebulae whose diameters are measured in hundreds of thousands of millions of miles." He tries to help out with the following: "If the Sun were a speck of dust 1⁄300 of an inch in diameter, it [that is, the speck-sized Sun] would have to extend 4 million miles in every direction to encompass even a few neighboring galaxies."

And also: "Empty Waterloo Station of everything except six specks of dust, and it is still far more crowded with dust than space is with stars."

And also: "The number of molecules in a pint of water placed end to end . . . would form a chain capable of encircling the Earth over 200 million times."

And, finally, he offers a way to imagine the stupendous heat involved in nuclear fusion. A pinhead heated to the temperature of the center of the Sun, writes Jeans, "would emit enough heat to kill anyone who ventured within a thousand miles of it."

These images carry emotional lessons that numbers alone cannot. They give us a sense—as well as knowledge—of what truly large numbers are about.

One of the main reasons that large numbers grow so explosively is that multiplication is a powerful engine for growth—even when the only number you happen to be multiplying is insignificantly puny, like the number two.

There's an old legend about the mathematician who invented chess that illustrates this very well. The king liked the game so well, the story goes, that he offered the mathematician any prize she wanted. She asked only for two grains of wheat to be placed on the first square of the chessboard, four on the second, eight on the third . . . and so forth, doubling the number of grains of wheat for each of the sixty-four squares on the chessboard.

How much grain did she win? More than human beings had produced in the entire history of the world. That's the power of doubling.

An even more vivid story comes from physicist Albert A. Bartlett, who has launched a one-person crusade in support of exponential literacy. The recipient of a Distinguished Service citation from the American Association of Physics teachers, Bartlett is currently professor emeritus at the University of Colorado, Boulder.

Here's the story he uses to demonstrate the precarious state of our natural resources, even in times of apparent wealth:

Imagine an average colony of bacteria, living in Bacterialand. They go off to found a new colony—in a Coke bottle found buried in the earth. They excavate it and make themselves at home. Let's say we start with two intrepid explorers who settle this new colony. Let's say they double their population once each minute. Let's say they start at eleven A.M., and by twelve noon their bottle is full, and they're out of space and resources.

What time would it be, Bartlett asks, when even the most far-sighted bacteria saw an overpopulation problem on the horizon? Certainly not before 11:58, he answers, because at that point the bottle would only be one-quarter full. (Two doublings away from full.) Even at 11:59, it would be only half full, and you can just hear the bacteria politicians singing platitudes like: No need to worry, folks! WE HAVE MORE SPACE LEFT IN OUR HOMELAND THAN WE'VE USED IN ALL THE HISTORY OF THIS COLONY!

Nevertheless, they decide to explore offshore for more Coke bottles. They find three! Wow! How long does this give the bacteria colony before it runs out of room again? Answer: Two minutes.

In fact, everything that grows exponentially doubles sooner or later. Compound interest at 7 percent doubles your money in ten years. Population growth at 7 percent doubles the number of people in ten years. Let's say we call the first two bacteria in the bottle Adam and Eve. The human population is expected to double in fifty to sixty years (by the most optimistic calculations). But it's unlikely that people will get too perturbed about it anytime soon, because the bottle appears to be still far from full—especially if you live in Montana.

Exponential growth is good news, of course, for people with money in the bank, or better, the stock market. One lousy dollar invested at a return of 5 percent a year compounded continually for five hundred years will flow in at the rate of $114 *per second.*

It's bad news for people with fixed incomes, like people on pensions or salaries that don't increase. It explains why a typical market basket of goods that cost $100 in 1982 cost 50 percent more—$142—in 1992. The same basket back in 1946 cost $19.50.

It's important to note that these numbers add up so fast only if the gains are compounded. That is, every time the population of bacteria doubled, the number that was multiplied by two was the sum total of all the previous multiplications. So what's doubling is rapidly growing, and the growth gets bigger the bigger it gets.

Compare, for example, the interest you'd earn if you put $1,000 in the bank at 10 percent interest for one hundred years. If you simply added $100 in interest each year, after a hundred years you'd have $11,000. But if you take 10 percent of the total principal plus interest each year, you wind up with more than $22 million.

These numbers creep up on us unawares in part because of the way our brains are engineered. Our minds don't perceive the explosive magnitudes of exponential amplification because our brains appear to be calibrated like the Richter scale,* which measures the power of earthquakes.

As most people know, a quake that registers 8 on the Richter scale is hugely more powerful (in fact, 10 times more powerful) than one that registers 7—a lot more than the difference between the numbers seven and eight would suggest. Richter scales work like the chessboard—where the energy of the quake goes up like the grains of wheat, but the number on the scale only goes up like the number of the square the grain sits on.

The same is true of neurons, or perhaps neural connections, in the sense organs and brain. The human eye can see a range of well over a million different shades of brightness—but we don't perceive the brightest thing we can see as a million times brighter than the dimmest. There simply isn't enough room in the brain. So sight works on a Richter scale.

*Actually, these days seismologists use more sophisticated versions of the familiar Richter scale—based on the same mathematical principles.

So does sound. "The intensity of sounds ranges over such an enormous range that there's no way you could get a linear [nonexponential] system to handle that," says Bartlett. "It's something nature's built into our hearing and eyesight."

Chances are, it's built into our counting ability, as well. Otherwise, a finite brain could not handle the exponential amplification that drives everything from the size scales of the universe to the growth of national budgets.

Certainly, exponential* scales play a part in human perception of time. We remember last week more clearly than the week before, and the week before much better than two weeks before. At the same time, each year makes up a smaller fraction of your life—so that one year at the age of two is half your total life span; at fifty, a year is merely a fiftieth—and seems to fly by.

As physicist Philip Morrison has said, all the senses play some part in presenting the world to us in this same "distorted" way. It could not be otherwise, because only a logarithmic scale can encompass such a huge range of responses.

We also don't perceive changes that are too gradual, too finely tuned, for our perceptual system to pick up.† We don't see mountains move or flowers grow, even though we know that some mountain peaks were once at the bottom of the sea and that flowers begin as tiny seeds. We'd need to speed up time enormously to perceive it. We don't see night fall, although we know that the day ends and darkness descends.

At the same time, we'd need to slow down time enormously to perceive most chemical reactions taking place in our brains (or anywhere else for that matter). This becomes relevant to our perception of exponential amplification because sometimes small changes are all

*Actually, the scale is logarithmic; the growth is exponential.

†See Chapter 6, "Emerging Properties: More Is Different."

we have to go on. We might need to see extremely fine-grained differences to determine just when, say, the Coke bottle is one-quarter full, when catastrophe-size climate changes are upon us.

Curiously, the human brain itself is a product of exponential amplification. According to Anthony Smith in his book *The Mind,* the human brain contains 10 to 15 billion nerve cells—three times as many as there are humans on the planet. If you add in the number of connections between nerve cells, the sum is more than the number of humans who ever lived. Fifteen billion is also more or less the number of stars in the galaxy.

The brain built up all this marvelous gray matter by doubling. To get to 15 billion nerve cells takes only thirty-three doublings of the first cell; to get to half that number takes just thirty-two doublings—which is about the size of the number of cells in the brain of an ape. Our brains are only one doubling away from our simian relatives.

Why does any of this matter? Think: inflation, epidemics, energy consumption, nuclear explosions, population explosions—you get the idea.

Take population growth. People have known since Malthus that population increases exponentially. But no one worries much because (some people argue) populations are self-limiting. When people run out of food and places to live, they die or go to war or stop having children or some combination of all of these. All this is true—to some extent.

Like the bacteria in the bottle, however, we're faced with the inability to look far enough ahead to stop catastrophe before it strikes. Looking ahead is especially hard to do when everything seems to be under control. Cornell University ecologist David Pimentel warned at the February 1996 meeting of the American Association for the Advancement of Science that "we're adding a quarter million people every twenty-four hours. . . . And no one is really taking action about it. It's not a big bang kind of problem; it's a gradual thing. But that trickle is going to nickel-and-dime us to death."

A closely connected problem is consumption. It's a given in virtually all economic discussions that growth is a desirable goal. To bring Third World countries into the affluence of modernity, we urge them to adopt U.S.-style consumption. It doesn't take a bacterium in a bottle to see where this leads.

And along with consumption comes waste. Some recently prominent magazine articles have have argued that recycling was hurting the environment (which may be true), but also argued that we could easily dispense with efforts to recycle glass and paper, since the United States still had ample places to stash its garbage. One wonders how many doublings of garbage will take place before somebody notices that Earth is a sphere—and therefore finite—surface.

Bartlett recently wrote a paper designating people who believed in the continual possibility of growth as the modern-day Flat Earth Society. That's because a flat Earth could extend infinitely in all directions, accommodating any amount of garbage or providing any amount of land to grow crops or atmosphere to absorb the gases we pump in.

But, alas, Earth is a sphere. It has no place to grow. What we have is all there is. That's not a political statement. It's simply mathematics. That's why the wildly popular term "sustainable growth" is an oxymoron, says Bartlett. No growth is sustainable on a spherical Earth. If the population is growing at only 1.9 percent a year, it will still double in thirty-six years. No matter how small you make the growth rate, you still get doubling.*

As the bacteria in the bottle so aptly illustrate, growth is never sustainable—whether it's population increase or resource consumption.

*Conversely, even a small decline in population—if it stays at a steady rate—will multiply exponentially, eventually leading to extinction. One can imagine a sustainable growth rate as one that would rise and fall in gentle cycles, hovering around zero.

Astrophysicist Joel Primack thinks science may come to the rescue in a rather surprising way. The moral principles that guide our society, he argues, have always been based to some extent on our understanding of the natural world. Every culture and religion, for example, has a creation story—an explanation of how and why the universe was created and how humans and other beings fit into the grand scheme. And whether we are aware of it or not, we base a great deal of our thinking and planning on the knowledge conveyed in these stories.

Now that cosmologists are beginning to understand creation in a literal sense—the big bang—a new era may well be upon us, and just in time. If today's cosmology is right—and most researchers suspect that it probably is—then the universe itself went through an explosive inflationary period soon after the big bang. When inflation cooled down, the universe continued to expand at its present stately pace.

Perhaps, Primack suggests, we will learn to take the universe as our role model. "Inflation is a taste of what it is like to be God," he argues. "It cannot be considered a normal human pace. In a finite environment, inflation cannot continue, however cleverly we may postpone or disguise the inevitable. This is a consequence of natural laws."

By rethinking our creation stories in light of current research, we may find a solution to some of humanity's most pressing problems.

Chapter 3

CALCULATED RISKS

Newsweek magazine plunged American women into a state of near panic some years ago when it announced that the chances of a college-educated thirty-five-year-old woman finding a husband was less than her chance of being killed by a terrorist. Although Susan Faludi made mincemeat of this so-called statistic in her book *Backlash,* the notion that we can precisely quantify risk has a strong hold on the Western psyche. Scientists, statisticians, and policy makers attach numbers to the risk of getting breast cancer or AIDS, to flying and food additives, to getting hit by lightning or falling in the bathtub.

Yet despite (or perhaps because of) all the numbers floating around, most people are quite properly confused about risk. I know people who live happily on the San Andreas Fault and yet are afraid to ride the New York subways (and vice versa). I've known smokers who can't stand to be in the same room with a fatty steak, and women afraid of the side effects of birth control pills who have unprotected sex with strangers. Risk assessment is rarely based on purely rational considerations—even if people could agree on what those considerations were. We worry about negligible quantities of

Alar in apples, yet shrug off the much higher probability of dying from smoking. We worry about flying, but not driving. We worry about getting brain cancer from cellular phones, although the link is quite tenuous. In fact, it's easy to make a statistical argument—albeit a fallacious one—that cellular phones prevent cancer, because the proportion of people with brain tumors is smaller among cell phone users than among the general population.*

Even simple pleasures such as eating and breathing have become suspect. Love has always been risky, and AIDS has made intimacy more perilous than ever. On the other hand, not having relationships may be riskier still. According to at least one study, the average male faces three times the threat of early death associated with not being married as he does from cancer.

Of course, risk isn't all bad. Without knowingly taking risks, no one would ever walk out the door, much less go to school, drive a car, have a baby, submit a proposal for a research grant, fall in love, or swim in the ocean. It's hard to have any fun, accomplish anything productive, or experience life without taking on risks—sometimes substantial ones. Life, after all, is a fatal disease, and the mortality rate for humans, at the end of the day, is 100 percent.

Yet, people are notoriously bad at risk assessment. I couldn't get over this feeling watching the aftermath of the crash of TWA Flight 800 and the horror it spread about flying, with the long lines at airports, the increased security measures, the stories about grieving families day after day in the newspaper, the ongoing attempt to figure out why and who and what could be done to prevent such a tragedy from happening again.

Meanwhile, tens of thousands of children die every day around the world from common causes such as malnutrition and disease. That's roughly the same as a hundred exploding jumbo jets full of children

*John Allen Paulos was the first person I know of to make this calculation; it is probably related to the fact that people who use cellular phones are on average richer, and therefore healthier, than people who don't.

every single day. People who care more about the victims of Flight 800 aren't callous or ignorant. It's just the way our minds work. Certain kinds of tragedies make an impact; others don't. Our perceptual apparatus is geared toward threats that are exotic, personal, erratic, and dramatic. This doesn't mean we're ignorant; just human.

This skewed perception of risk has serious social consequences, however. We aim our resources at phantoms, while real hazards are ignored. Parents, for example, tend to rate drug abuse and abduction by strangers as the greatest threats to their children. Yet hundreds of times more children die each year from choking, burns, falls, drowning, and other accidents that public safety efforts generally ignore.

We spend millions to fight international terrorism and wear combat fatigues for a morning walk to protect against Lyme disease. At the same time, "we see several very major problems that have received relatively little attention," write Bernard Cohen and I-Sing Lee in *Health Physics.* The physicists suggest—not entirely tongue in cheek—that resources might be far more efficiently spent on programs such as government-organized computer dating services. "Favorable publicity on the advantages of marriage might be encouraged."

It's as if we incarcerated every petty criminal with zeal, while inviting mass murderers into our bedrooms. If we wanted to put the money on the real killers, we'd go after suicide, not asbestos.

Even in terms of simple dollars, our policies don't make any sense. It's well known, for example, that prenatal care for pregnant women saves enormous amounts of money—in terms of care infants need in the first year of life—and costs a pittance. Yet millions of low-income women don't get it.

Numbers are clearly not enough to make sense of risk assessment. Context counts, too. Take cancer statistics. It's always frightening to hear that cancer is on the rise. However, at least one reason for the increase is simply that people are living longer—long enough to get the disease.

Certain conclusions we draw from statistics are downright silly. Physicist Hal Lewis writes in *Technological Risk* that per mile traveled a person is more likely to be killed by a car as a pedestrian than as a driver or passenger. Should we conclude that driving is safer than walking and therefore that all pedestrians should be forced into cars?

Charles Dickens made a point about the absurdity of misunderstanding numbers associated with risk by refusing to ride the train. One day late in December, the story goes, Dickens announced that he couldn't travel by train any more that year, "on the grounds that the average annual quota of railroad accidents in Britain had not been filled and therefore further disasters were obviously imminent."

Purely numerical comparisons also may be socially unacceptable. When the state of Oregon decided to rank its medical services according to benefit-cost ratios, some results had to be thrown out—despite their statistical validity. Treatment for thumb sucking, crooked teeth, and headaches, for example, came out on the priorities list ahead of therapy for cystic fibrosis and AIDS.

What you consider risky, after all, depends somewhat on the circumstances of your life and lifestyle. People who don't have enough to eat don't worry about apples contaminated with Alar. People who face daily violence at their front door don't worry about hijackings on flights to the Bahamas. Attitudes toward risk evolve in cultural contexts and are influenced by everything from psychology to ethics to beliefs about personal responsibility.

In addition to context, another factor needed to see through the maze of conflicting messages about risk is human psychology. For example, imminent risks strike much more fear in our hearts than distant ones; it's much harder to get a teenager than an older person to take long-term dangers like smoking seriously.

Smoking is also a habit people believe they can control, which makes the risk far more acceptable. (People seem to get more upset about the effects of passive smoking than smoking itself—at least in part because smokers get to choose, and breathers don't.)

As a general principle, people tend to grossly exaggerate the risk of any danger perceived to be beyond their control, while shrugging off risks they think they can manage. Thus, we go skiing and skydiving, but fear asbestos. We resent and fear the idea that anonymous chemical companies are putting additives into our food; yet the additives we load onto our own food—salt, sugar, butter—are millions of times more dangerous.

This is one reason that airline accidents seem so unacceptable—because strapped into our seats in the cabin, what happens is completely beyond our control. In a poll taken soon after the TWA Flight 800 crash, an overwhelming majority of people said they'd be willing to pay up to fifty dollars more for a round-trip ticket if it increased airline safety. Yet the same people resist moves to improve automobile safety, for example, especially if it costs money.

The idea that we can control what happens also influences who we blame when things go wrong. Most people don't like to pay the costs for treating people injured by cigarettes or riding motorcycles because we think they brought these things on themselves. Some people also hold these attitudes toward victims of AIDS, or mental illness, because they think the illness results from lack of character or personal morals.

In another curious perceptual twist, risks associated with losing something and gaining something appear to be calculated in our minds according to quite different scales. In a now-classic series of studies, Stanford psychologist Amos Tversky and colleague Daniel Kahneman concluded that most people will bend over backward to avoid small risks, even if that means sacrificing great potential rewards. "The threat of a loss has a greater impact on a decision than the possibility of an equivalent gain," they concluded.

In one of their tests, Tversky and Kahneman asked physicians to choose between two strategies for combating a rare disease, expected to kill 600 people. Strategy A promised to save 200 people (the rest would die), while Strategy B offered a one-third probability that everyone would be saved, and a two-thirds probability that no one would be saved. Betting on a sure thing, the physicians choose A. But presented with the identical choice, stated differently, they choose B. The difference in language was simply this: Instead of stating that Strategy A would guarantee 200 out of 600 saved lives, it stated that Strategy A would mean 400 sure deaths.

People will risk a lot to prevent a loss, in other words, but risk very little for possible gain. Running into a burning house to save a pet or fighting back when a mugger asks for your wallet are both high-risk gambles that people take repeatedly in order to hang on to something they care about. The same people might not risk the hassle of, say, fastening a seat belt in a car even though the potential gain might be much higher.

The bird in the hand always seems more attractive than the two in the bush. Even if holding on to the one in your hand comes at a higher risk and the two in the bush are gold-plated.

The reverse situation comes into play when we judge risks of commission versus risks of omission. A risk that you assume by actually doing something seems far more risky than a risk you take by not doing something, even though the risk of doing nothing may be greater.

Death from natural causes, like cancer, are more readily acceptable than deaths from accidents or murder. That's probably one reason it's so much easier to accept thousands of starving children than the death of one in a drive-by shooting. The former is an act of omission—a failure to step in and help, send food or medicine. The latter is the commission of a crime—somebody pulled the trigger.

In the same way, the Food and Drug Administration is far more likely to withhold a drug that might help a great number of people if it threatens to harm a few; better to hurt a lot of people by failing to do something than act with the deliberate knowledge that some people will be hurt. Or as the doctors' credo puts it: First do no harm.

For obvious reasons, dramatic or exotic risks seem far more dangerous than more familiar ones. Plane crashes and AIDS are risks associated with ambulances and flashing lights, sex and drugs. While red dye #2 strikes terror in our hearts, that great glob of butter melting into our baked potato is accepted as an old friend. "A woman drives down the street with her child romping around in the front seat," says John Allen Paulos. "Then they arrive at the shopping mall, and she grabs the child's hand so hard it hurts, because she's afraid he'll be kidnapped."

Children who are kidnapped are far more likely to be whisked away by relatives than strangers, just as most people are murdered by people they know.

Familiar risks creep up on us like age and are often difficult to see until it's too late to take action. Mathematician Sam C. Saunders of Washington State University reminds us that a frog placed in hot water will struggle to escape, but the same frog placed in cool water that's slowly warmed up will sit peacefully until it's cooked. "One cannot anticipate what one does not perceive," he says, which is why gradual accumulations of risk due to lifestyle choices (like smoking or eating) are so often ignored. We're in hot water, but it's gotten hot so slowly that no one notices.

To bring home his point, Saunders asks us to imagine that cigarettes are not harmful—with the exception of an occasional one that has been packed with explosives instead of tobacco. These dynamite-stuffed cigarettes look just like normal ones. There's only one hidden away in every 18,250 packs—not a grave risk, you might say. The only catch is, if you smoke one of those explosive cigarettes, it might blow your head off.

The mathematician speculates, I think correctly, that given such a situation, cigarettes would surely be banned outright. After all, if 30 million packs of cigarettes are sold each day, an average of 1,600 people a day would die in gruesome explosions. Yet the number of deaths is the same to be expected from normal smoking. "The total expected loss of life or health to smokers using dynamite-loaded (but otherwise harmless) cigarettes over forty years would not be as great as with ordinary filtered cigarettes," says Saunders.

We can accept getting cooked like a frog, in other words, but not getting blown up like a firecracker.

It won't come as a great surprise to anyone that ego also plays a role in the way we assess risks. Psychological self-protection leads us to draw consistently wrong conclusions. In general, we overestimate the risks of bad things happening to others, while vastly underrating the possibility that they will happen to ourselves. Indeed, the lengths people go to minimize their own perceived risks can be downright "ingenious," according to Rutgers psychologist Neil Weinstein. For example, people asked about the risk of finding radon in their houses always rate their risk as "low" or "average," never "high." "If you ask them why," says Weinstein, "they take anything and twist it around in a way that reassures them. Some say their risk is low because the house is new; others, because the house is old. Some will say their risk is low because their house is at the top of a hill; others, because it's at the bottom of a hill."

Whatever the evidence to the contrary, we think: "It won't happen to me." Weinstein and others speculate that this has something to do with preservation of self-esteem. We don't like to see ourselves as vulnerable. We like to think we've got some magical edge over the others. Ego gets involved especially in cases where being vulnerable to risk implies personal failure—for example, the risk of depression, suicide, alcoholism, drug addiction. "If you admit you're at risk," says Weinstein, "you're admitting that you can't handle stress. You're not as strong as the next person."

Average people, studies have shown, believe that they will enjoy longer lives, healthier lives, and longer marriages than the "average" person. Despite the obvious fact that they themselves are, well, average people, too. According to a recent poll, 3 out of 4 baby boomers (those born between 1946 and 1964) think they look younger than their peers, and 4 out of 5 say they have fewer wrinkles than other people their age—a statistical impossibility.

Kahneman and Tversky studied this phenomenon as well and found that people think they'll beat the odds because they're special. This is no doubt a necessary psychological defense mechanism, or no one would ever get married again without thinking seriously about the potential for divorce. A clear view of personal vulnerability, however, could go a long way toward preventing activities like drunken driving. But then again, most people think they are better than average drivers—even when intoxicated.

We also seem to believe it won't happen to us if it hasn't happened yet. That is, we extrapolate from the past to the future. "I've been taking that highway at eighty miles per hour for ten years and I haven't crashed yet," we tell ourselves. This is rather like reasoning that flipping a coin ten times that comes up heads guarantees that heads will continue to come up indefinitely.*

Curiously, one advertising campaign against drunken driving that was quite successful featured the faces of children killed by

*See Probable Causes in Chapter 12, "The Truth about Why Things Happen."

drunken drivers. These children looked real to us. We could identify with them. In the same way as we could identify with the people on TWA Flight 800. It's much easier to empathize with someone who has a name and a face than a statistic.

That explains in part why we go to great expense to rescue children who fall down mine shafts, but not children dying from preventable diseases. Economists call this the "rule of rescue." If you know that someone is in danger and you know that you can help, you have a moral obligation to do so. If you don't know about it, however, you have no obligation. Columnist Roger Simon speculates that's one reason the National Rifle Association lobbied successfully to eliminate the program at the Centers for Disease Control that keeps track of gun deaths. If we don't have to face what's happening, we won't feel obligated to do anything about it.

Even without the complication of all these psychological factors, however, calculating risks can be tricky because not everything is known about every situation. "We have to concede that a single neglected or unrecognized risk can invalidate all the reliability calculations, which are based on known risk," writes Ivar Ekeland. There is always a risk, in other words, that the risk assessment itself is wrong.

Genetic screening, like tests for HIV infection, has a certain probability of being wrong. If your results come back positive, how much should you worry? If they come back negative, how safe should you feel?

The more factors involved, the more complicated the risk assessment becomes. When you get to truly complex systems like nationwide telephone networks and power grids, worldwide computer networks and hugely complex machines like space shuttles, the risk of disaster becomes infinitely harder to pin down. No one knows when a minor glitch will set off a chain reaction of events that will culminate in disaster. Potential risk in complex systems, in other words, are subject to the same kinds of exponential amplification discussed in the previous chapter.

Needless to say, the way a society assesses risk is very different from the way an individual views the same choices. Whether or not you wish to ride a motorcycle is your own business. Whether society pays the bills for the thousands of people maimed by cycle accidents, however, is everybody's business. Any one of us might view our own survival on a transatlantic flight as more important than the needs of the nation's children. Governments, one presumes, ought to have a somewhat different agenda.

But how far does society want to go in strictly numerical accounting? It certainly hasn't helped much in the all-important issue of health care, where an ounce of prevention has been proven again and again to be worth many pounds of cures. Most experts agree that we should be spending much more money preventing common diseases and accidents, especially in children. But no one wants to take health dollars away from precarious newborns or the elderly—where most of it goes. These are decisions that ultimately will not be made by numbers alone. Calculating risk only helps us to see more clearly what exactly is going on.

According to anthropologist Melvin Konner, author of *Why the Reckless Survive,* our poor judgment about potential risks may well be the legacy of evolution. Early peoples lived at constant risk from predators, disease, accidents. They died young. And in evolutionary terms, "winning" means not longevity, but merely sticking around long enough to pass on your genes to the next generation. Taking risk was therefore a "winning" strategy, especially if it meant a chance to mate before dying. Besides, decisions had to be made quickly. If going for a meal of ripe berries meant risking an attack from a saber-toothed tiger, you dove for the berries. For a half-starved cave dweller, this was a relatively simple choice. Perhaps our brains are simply not wired, speculates Konner, for the careful calculations presented by the risks of modern life.

Indeed, some of our optimistic biases toward personal risk may still serve important psychological purposes. In times of stress and danger, they help us to put one foot in front of the other; they help us to get on with our lives, and out the door.

In the end, Konner, the cautious professor, ruminates somewhat wistfully about his risk-taking friends—who smoke, and ride motorcycles, and drive with their seat belts fastened behind them. Beside them, he feels "safe and virtuous," yet somehow uneasy. "I sometimes think," he muses, "that the more reckless among us may have something to teach the careful about the sort of immortality that comes from living fully every day."

PART II

Interpreting the Physical World

Mountains are not pyramids and trees are not cones.
God must love gunnery and architecture if Euclid is his only geometer.
—*Thomasina, in Tom Stoppard's* Arcadia

Looking for truth in numbers presents obstacles far beyond the peculiar nature of the human thinking apparatus we carry around in our heads (and also in the rest of our bodies). There's also the difficulty of getting true information from what some people call the real world. We only glimpse that real world through the patterns, or signals, we see in our heads. But those patterns and signals are created, at least in part, outside ourselves: Call them information, messages, signals, relationships, ideas. Whatever you call them (how's input for a modern, computer-related analogy?), understanding anything requires getting a handle on that stuff out there and on the ways in which knowledge about it arrives on our internal radar screens.

Needless to say, the relationship between human senses and the world outside is a subject of breathtaking enormity—comprising the efforts of religion, science, philosophy, and art over many centuries. The few aspects discussed here offer just a taste of some of the glitches we may encounter in trying to add up what's going on out there, mathematically speaking.

For example, how do we measure the properties of things? How do the quantities we measure with scales and thermometers and IQ tests relate to qualities? Most of us have fallen into the long-practiced habit of believing that quality and quantity are two different kinds of properties that don't affect each other. On the contrary, not only does quantity frequently determine quality, quantity (or, more broadly, scale) can affect the very notion of what is true, what is possible, what, indeed, exists.

Another limit imposed by reality is its sheer complexity, which makes it impossible to predict some ordinary things (like weather) at the same time that it's possible to predict truly extraordinary things (like the fate of the universe). Astronomers can measure the distance from Earth to the Sun to the nearest centimeter. But they can't predict the time of the sunset to the nearest minute—because at least nine different factors are involved. (Some of the most critical variables are the speed and tilt of Earth's orbit and the layers of smog that can bend the setting rays of light like a lens.)

Yet another curious aspect of the information we glean from the outside world is that one person's data is another person's noise, and knowing which is which in any particular instance is not a simple matter. Indeed, what's noise today may be viewed as an important signal tomorrow, and vice versa. Like so much else about information, the difference between signals and noise frequently depends entirely on context.

Chapter 4

THE MEASURE OF MAN, WOMAN, AND THING

They ask wavy questions to decide whether it's a wave,
and particley questions to decide whether it's a particle.
—*Ian Stewart and Jack Cohen,* The Collapse of Chaos

How do I love thee? Let me count the ways. The Vatican measures saintliness, the military courage, the judicial system penitence, dog breeders temperament. Both physicians and policy makers measure human worth. Juries put dollars and cents on pain and suffering. If you want to get to the truth about something, the first thing you do is size it up. To get a handle on the economy or human nature or the behavior of subatomic particles or the birth of stars or the well-being of a patient, you need to start with data—and data comes from making careful measurements.

Measurement, it's probably fair to say, is the cornerstone of knowledge. It allows us to compare things with other things and to quantify relationships. Mine is bigger than yours amounts to a mathematical statement ($M > Y$). My country/child/car is better/smarter/faster than anyone else's. As a society, we take these measurements extremely seriously. Indeed, during the 1996 Summer Olympics, it was disconcerting to see how the grace of a gymnastics routine or the precision of a dive could be distilled into three decimal places.

We put numerical measures onto everything from public opinion to fat content, from wealth to waistlines, from nuclear reactions to success. Can all these disparate things be measured? Can quality be boiled down to quantity?

Ultimately, everything boils down to quantity, if only because everything ultimately boils down to elementary particles and forces. But such numerical accounting misses a huge range of phenomena that only emerges on the macroscopic scale—from Santa Claus to love. More substantively, anything you can define can be measured simply because the act of defining it also defines the qualities that make it up.

But measurements have meaning only to the extent that we respect their various limits. No measurements, when you get right down to it, are straightforward. All involve disentangling things that can't be separated, or quantifying things that can't be counted, or defining things you can't quite put your finger on. Usually, the act of measuring something affects it; sometimes, measurement destroys it.

Ultimately, asking questions of nature is always somewhat treacherous because the answers you get depend on the questions you ask—even when the questions you're asking are as innocent as "How much?" If you stand on a scale, you will not get any information about your temperature. You can't use a tape measure to gauge the air pressure in your tires, or a watch to measure musicality.

The answers also depend on where you (the observer) are standing. A different frame of reference can yield wildly different answers on questions of space and time, for example.*

*See Shifting Frames in Chapter 14, "Emmy and Albert."

Indeed, neither space nor time can be measured in isolation from the other. You might look up one night and see a faint blur in the sky that is the Andromeda galaxy—our nearest spiral neighbor, more than 2 million light-years away. The light that reaches your retina now left Andromeda more than 2 million years ago, long before Homo sapiens walked on Earth. So the Andromeda we see today is really old (one might say ancient) news, 2 million years stale. But suppose you want to know what's happening on Andromeda *now.*

As it turns out, the question has no meaning, because there is no way of measuring a simultaneous "now" in the universe. We can't synchronize our watches because nothing can travel faster than light. And the light that's reached us from Andromeda has been traveling as fast as it can, at the speed limit of the universe. In the only sense that makes sense, we *are* seeing Andromeda now.

Space and time, in other words, are inseparable partners. You can talk about here and now, and there and then. But it makes no sense to talk about there and now. Even if you stand in one place on Earth and measure the time now, and again two seconds from now, you have also measured motion in space—for the simple reason that the Earth moves. If it didn't, we wouldn't have such a thing as a day to be divided into hours and minutes and seconds in the first place. In a way, measuring the distance that a point on the surface of Earth moves in a given moment of time can be viewed as a tautology, since a moment of time is defined by the rotation of Earth through space. If nothing ever happened in the universe, there would be no conceivable concept of time.

Other quantities we try to measure are equally entangled. Matter and energy, for example, or mind and brain, or the influence of genetic and environmental factors on intelligence.

But even if such factors could be disentangled (or deconvoluted, as the mathematicians like to say), it's not clear they could be readily measured. Even precisely defined and seemingly isolated quantities have a way of slipping out of our attempts to pin them down, like the proverbial greased pigs.

For example, subatomic particles cannot be precisely measured without making quantifiable sacrifices. If you measure precisely what a particle is doing, you cannot at the same time measure precisely where it is. If you measure precisely how much energy it has, you lose all information about time.

Subatomic physics gives some people the willies because of this inherent uncertainty. But knowledge is also lost with the most mundane measurements of everyday life. You cannot chemically analyze your dinner and eat it (or at least the very same bit), too. You cannot dissect the mathematics underlying Mozart and at the same time feel the emotional impact. A Picasso, viewed through a powerful microscope, dissolves into a grainy pattern of dots. Earth, viewed from space, reveals itself as a sphere but tells you nothing about what's going on in your own backyard. Something is lost for every measurement that's gained. If you test a tree for its boatlike features, writes Ian Stewart, you can't simultaneously test how good it will be for holding up telephone lines.

According to quantum mechanics, your very choice to measure something affects—and even determines—the measurement you make. The most famous example of this paradoxical proposition is a thought experiment concocted by Erwin Schrödinger to expose what he felt were the obvious absurdities of quantum theory. If you followed the reasoning of quantum theory to its logical conclusions, he said, you could have a cat sitting in a closed box, which was both dead and alive at the same time. The cat would only become wholly alive or dead when you opened the box. Therefore, the very act of opening the box to observe the cat would either kill it or save its life.

The experiment goes like this: You put a cat in box with a glass vial of poison. A hammer capable of breaking the vial (and therefore killing the cat) is attached to a gizmo that can be set off by the radioactive decay of a certain element. The probability that the atom will decay—therefore setting off the gizmo and killing the cat—is exactly fifty-fifty. You have no way of knowing if it has decayed until you open the box. Ergo, opening the box either kills or saves the cat.*

A simpler way to look at this, as Stewart suggests, is to imagine a spinning coin. As it spins, it is neither heads nor tails, but some combination of both. However, in order to measure whether it's heads or tails, you have to stop the coin from spinning—therefore forcing it to choose heads or tails.†

Measuring one aspect, in other words, destroys the ability to measure another. Again, such a situation is not without precedent in everyday life. The stress of test taking, for example, can so alter some people's ability to think clearly that the test ends up measuring only stress, not knowledge or aptitude. Reporters given the task of accurately and objectively representing events frequently influence those events and occasionally create them out of whole cloth (whether they wish to or not).

The answer is determined not only by the type of question, but also on the very fact that there *is* a question.

However one approaches measurement, at some point the process is going to involve coming up with a number to describe something—whether that something is describable in numbers or not. While this may be obvious when it comes to human qualities, it's equally relevant in certain realms of so-called hard science. Take the numbers

*Cat owners like myself may be puzzled at the notion that a half-dead cat presents a paradox, since cats seem quite at home in a half-dead state. A more substantive objection to the paradox is that cats are not simple quantum mechanical objects and so shouldn't be expected to behave in simple quantum mechanical ways. See Chapter 6, "Emerging Properties."

†This is obviously an oversimplification. If quantum theory were that simple, physicists wouldn't still be arguing about its meaning. However, this analogy does help nonquantum-mechanically inclined readers to get a handle on the concept.

physicists use to describe subatomic particles, for example. These quantum numbers refer to certain quantities that, put together, give the particle its identity. For example, a particle can have an electric charge of 1, minus 1, or 0 (subatomic particles called quarks can carry electric charge in multiples of ⅓; molecules can carry plus or minus 2 or more). Particles also carry a certain amount of a property called spin—again, in integer (or half-integer) amounts.

Do these measures really describe the atom? In some sense, yes. Researchers can use these numbers to combine atoms in various ways and predict their properties quite accurately—and even invent entirely new kinds of materials from scratch.

But in another sense, the understanding is only an approximation. Atoms aren't numbers. Or miniature solar systems. They can only be truly described in their own unworldly terms. Or as physicist Arthur Eddington puts it: "In short, the physicist draws up an elaborate plan of the atom and then proceeds critically to erase each detail in turn. What is left is the atom of modern physics."

Still, quantum numbers describe atoms a whole lot better than, say, IQ scores describe intelligence. That's partly because physicists have refined their definitions of atomic properties better than people have defined what it means to be smart (not to mention wise).

Other immeasurables aren't so obvious, though. For example, how do you measure success? To make it easy, let's just say economic success, and limit comparisons to countries. Is there any accurate way to judge which country is more economically successful than another?

On the face of it, this is a snap. You simply measure GNP (gross national product), the sum of all the goods and services the economy produces. But as Harvard University economist Amartya Sen has pointed out, traditional measures such as GNP neglect critical factors such as well-being. Famine often exists side by side with plentiful food, and high starvation rates are common in wealthy countries. Sen suggests mortality rates are a much better measure, because they more accurately reflect national well-being. By those measures, American blacks are worse off than the poor of many Third World countries.

Mathematician William Adams made much the same point using the economic boom of the 1980s as an example. During the Reagan years, the U.S. economy created $30 trillion worth of goods and services; employment grew; the stock market went ballistic. But by other measures, he argues, the boom was a disaster. Everything from the interstate highway system to public education was neglected almost to the point of destruction. So America ended the 1980s as the country with the most wealth—and also one of the most poorly educated populations in the Western world, with substandard health care and a deteriorating infrastructure.

Making a measurement also requires knowing exactly where the thing to be measured begins and ends. Since even few physical objects come marked with clear boundaries, this can be a close to hopeless task. Physicists who study clouds, for example, have a hard time getting a handle on their object of study. Where does a cloud begin? It may seem like a fluffy self-contained pillow from Earth, but fly through it in a plane and the edges evaporate into an indeterminate gauze of vapor.

Forced into a game of Trivial Pursuit by my family once, I couldn't answer a question that asked: How many colors are there in a rainbow? Since the spectrum is a continuous band of frequencies, the question is unanswerable—except, perhaps, as infinite. As Leslie White points out in Newman's *World of Mathematics:* One is inclined to think that yellow, blue, and green are features of the external world that any normal person would distinguish until learning that the Greeks and Natchez Indians did not distinguish yellow from green; they had one term for both.

One of the most egregious examples of drawing boundaries where none exist pops up in matters of race. Despite the fact that application forms and census forms ask people to choose whether they are black, white, Asian, or other, race is not a biological concept. In a recent survey, 41 percent of anthropologists said that there is no such biological thing as race.

Indeed, in a chemical sense, we really blend in with the people around us like spilled paint; your molecules and mine are continually drifting off the surface of our skin, breathed out of our noses, flaked off hair and scalps. Individuals don't have sharp edges. We blur into each other's space like perfume molecules wafting from an open bottle.

Our endless desire to put sharp edges on things, and then measure them precisely and cleanly, has led to problems even in simple math. No doubt, one of the reasons the Pythagoreans had such a pathological fear of irrational numbers stems from their inherent fuzziness. Ratios and integers have edges; a number is 2, or $3/4$, or $9/17$. But irrationals like pi or the square root of two can't be contained in boundaries. Like the Energizer Bunny, they just keep going and going and going. . . .

Perhaps the most insidious obstacle to accurate measurement is the normally unacknowledged fact that you can only measure the stuff that you go out and look for, stuff that you know (or suspect) is actually there. Astronomy is a humbling science in this regard, because it seems as if every time astronomers find a new way to measure the universe, they find an entirely new and unexpected category of species. Before the *Voyager* spacecraft visited Jupiter in the late 1970s, everyone assumed that Saturn was the only planet surrounded by rings. Almost as a lark, some researchers lobbied for an experiment to look for rings around Jupiter: Presto, there they were!

Until the middle of this century, astronomers could only look at things that glowed in visible light, and even with this rather limited perspective, revolutions were unleashed. Galileo's first primitive telescope was sharp enough to make the then radical discovery that other worlds had moons (Jupiter) and that celestial objects were not perfect, unblemished spheres (there were craters and mountains on the Moon). In a similar leap hundreds of years later, the hundred-inch Hooker telescope at Mount Wilson discovered that smudges seen in earlier telescopes were actually other island universes, galaxies just like our own. And as Carl Sagan reminded us, there are billions and billions of them!

And that was just the start. Today, astronomers see and hear the universe as it calls to them in radio, infrared, ultraviolet, X-ray, and gamma-ray radiation. Each time they tune in to a new station, they find something unexpected: pulsars and quasars and black holes. Who knows what the new generation of gravity wave telescopes will find. And who knows what the astronomers haven't even thought to look for yet. Almost as soon as their sights got sharp enough to see planets around other stars, they began to find them.

Each time they ask a new question, they peel off a new layer. No wonder Frank Oppenheimer used to call science the search for the ever juicier mystery.

Occasionally, someone will argue that astronomers have found everything there is; there are no more major surprises in store—at least not that we'll ever have the capability to detect and measure. When they say this, I'm always reminded of the French philosopher Auguste Comte, who stated baldly that if anything was absolutely certain, it was that people would never be able to measure the chemical composition of stars—which certainly seemed a safe enough pronouncement in 1825.

Before the end of the century, however, astronomers had learned to read starlight like words in a book. Since each atom absorbs and emits only particular frequencies, or colors, of light, each set of lines spells out the signature of a single kind of atom. Further, the signature changes depending on how much energy the atom has, what form it's in, and how it's moving.

Today we can measure the chemistry of the stars far better than we can measure the chemistry going on inside our own Earth—because stars are mostly transparent, and Earth is not.

There is even a counterpart to the astronomical experience in the subatomic realm. When physicists do experiments designed to detect waves, they find waves. When they do experiments designed to detect particles, they get particles. Normally, we don't have to deal with the evasive answers offered up by subatomic particles because we deal with large congregates of particles—and they behave, well, normally—that is to say, like waves or particles, but generally not both. Most things we measure are much bigger than atoms.

Frequently, however, congregating things into groups can make measurements even more deceptive. For example, since it often isn't possible (or manageable) to measure every individual within a group, we might substitute the group's average value. But averages meant to represent everybody in a group don't necessarily represent any real individual—as in the oft-quoted statistic: The average U.S. family has 2.5 children. When someone offers statistics stating, say, that boys on average do better in math than girls on average, that doesn't say anything at all about the abilities of any particular boy or girl.

Congregating multiple qualities of a single individual can be even more difficult than comparing groups. That is, it's easy enough to compare who's taller than who, or who's faster. But there's no sensible way to rank tallest and fastest together—that is, to find the tallest and fastest person in a group. It's even worse when you have to rank individuals, say job applicants, according to a whole range of qualities: responsibility, intelligence, thoughtfulness, knowledge, creativity, personality, and so forth.

As it turns out, there's no single solution. There is no way mathematically to well order more than two variables at the same time.

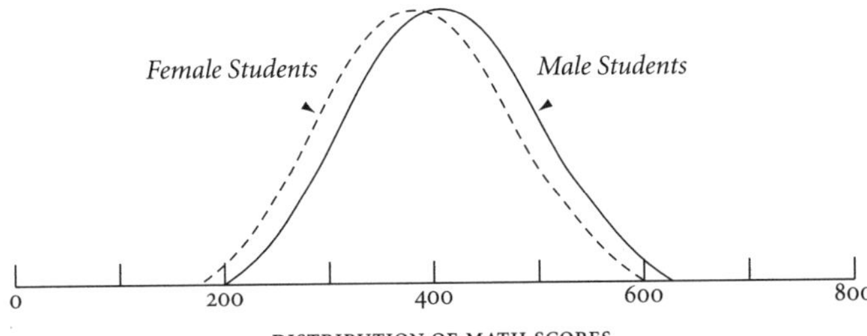

The two overlapping bell curves show how little difference a difference can make. Even though male students on average score higher than female students on average in this graphic, it's clear that a great number of females score far higher than a great many males—making it impossible to draw any conclusions about individual males or females from this information.

Think of a ranking as a point on a line. You can easily imagine the curve of the graph and how simple it would be to see what point has the higher rank. Now imagine adding another variable—in other words, making the line into a square. Now there is no way to well order any two points in the square, saying which comes before or after the next.

One way around this problem is to combine all the measurements into a single "figure of merit." UPS uses just such a system to rank irregularly shaped boxes. (Basically, they add the girth to the length—something like adding your height to your weight, a measure that makes no real sense, but may be useful nonetheless.) It's a way of adding up all the important attributes of something to arrive at a single number that can then be compared with other similar things.*

Presumably, you could come up with a figure of merit for anything from sexual attractiveness to chocolate mousse. But the ability

*The physicist and writer Hans Christian Von Baeyer used this example in an essay in *Discover* magazine to illustrate how a new ¾-mile-round accelerator going on-line in Virginia is actually more powerful than the huge seventeen-mile-round European Electron-Positron Collider—using a new figure of merit he invented specifically for the purpose.

of such a number to allow you to make useful comparisons would still depend on the judgments behind the numbers that went into it.

Finally, every measurement comes with what physicists call an error bar—a particularly graphic way to see right off what kind of probable error you're dealing with. In some cases, the error bar is larger than the measurement itself, which suggests that the measurement shouldn't be taken very seriously. Surprisingly, however, measurements with huge error bars frequently accompany scientific papers—and authors sometimes ignore the obvious implications in their conclusions.

Darrell Huff, in his classic book *How to Lie with Statistics*, gives the unfortunately common example of intelligence tests. An IQ test that gives one person a score of 98 (error of plus or minus 3) and another person a 101 (plus or minus 3) tells you absolutely nothing. What the numbers really mean is that there's a fifty-fifty chance that the person scoring 98 has an IQ that lies somewhere between 95 and 101. And there's a fifty-fifty chance that the person scoring 101 has an IQ between 98 and 104. That means the person scoring 98 could be superior to the person scoring 101 by three points.

As Huff reminds us: "A difference is a difference only if it makes a difference."

Or as Caltech physicist David Goodstein added when he read this section: "The very best physicists make this mistake when they give students tests, then take the scores seriously, as if there were any real difference between a 79 (C) and an 80 (B)."

Chapter 5

A MATTER OF SCALE

How would you suspend 500,000 pounds of water in the air with no visible means of support? (Answer: build a cloud.)

—*artist Bob Miller*

There is something magically seductive about an invitation to a world where everything measures much bigger or smaller than ourselves. To contemplate the vast expanse of ocean or sky, to look at pond scum under a microscope, to imagine the intimate inner life of atoms, all cast spells that take us far beyond the realm of everyday living into exotic landscapes accessible only through the imagination. What would it be like to grow as big as a giant? As small as a bug? Alice ate a mushroom and puffed up like a Macy's Thanksgiving Day balloon, bursting out of her house; she ate some more and shrank like the Incredible Shrinking Woman, forever in fear of falling down the drain. From Stuart Little to King Kong, from *Honey, I Shrunk the Kids* to Thumbelina, the notion of changing size seems to have a powerful pull on our psyches.

There are good reasons to think a world that's different in scale will also be different in kind. More or less of something very often adds up to more than simply more or less; quantitative changes can make huge qualitative differences.

When the size of things changes radically, different laws of nature rule, time ticks according to different clocks, new worlds appear out of nowhere while old ones dissolve into invisibility. Consider the strange situation of a giant, for example. Big and strong to be sure, but size comes with distinct disadvantages. According to J. B. S. Haldane in his classic essay, "On Being the Right Size," a sixty-foot giant would break his thighbones at every step. The reason is simple geometry. Height increases only in one dimension, area in two, volume in three. If you doubled the height of a man, the cross section, or thickness, of muscle that supports him against gravity would quadruple (two times two) and his volume—and therefore weight—would increase by a factor of eight. If you made him ten times taller, his weight would be a thousand times greater, but the cross section of bones and muscles to support him would only increase by a factor of one hundred. Result: shattered bones.

To bear such weight would require stout, thick legs—think elephant or rhino. Leaping would be out of the question. Superman must have been a flea.

Fleas, of course, perform superhuman feats routinely (which is part of the science behind the now nearly extinct art of the flea circus). These puny critters can pull 160,000 times their own weight, and jump a hundred times their own height. Small creatures have so little mass compared to the area of their muscles that they seem enormously strong. While their muscles are many orders of magnitude weaker than ours, the mass they have to push around is so much smaller that it makes each ant and flea into a superbeing. Leaping over tall buildings does not pose a problem.*

*According to Exploratorium physicist Tom Humphrey, all animals jump to the same height, roughly speaking. Both fleas and humans can jump about a meter off the ground—an interesting invariant. See Truth and Beauty in Chapter 14, "Emmy and Albert."

Neither does falling. The old saying is true: The bigger they come, the harder they fall. And the smaller they come, the softer their landings. Again, the reason is geometry. If an elephant falls from a building, gravity pulls strongly on its huge mass while its comparatively small surface area offers little resistance. A mouse, on the other hand, is so small in volume (and therefore mass) that gravity has little to attract; at the same time, its relative surface area is so huge that it serves as a built-in parachute.

A mouse, writes Haldane, could be dropped from a thousand-yard-high cliff and walk away unharmed. A rat would probably suffer enough damage to be killed. A person would certainly be killed. And a horse, he tells us, "splashes."

The same relationships apply to inanimate falling objects—say, drops of water. The atmosphere is drenched with water vapor, even when we can't see it in the form of clouds. However, once a tiny particle begins to attract water molecules to its sides, things change rapidly. As the diameter of the growing droplet increases by a hundred, the surface area increases by ten thousand, and its volume a millionfold. The larger surface area reflects far more light—making the cloud visible. The enormously increased volume gives the drops the gravitational pull they need to splash down to the ground as rain.

According to cloud experts, water droplets in the air are simultaneously pulled on by electrical forces of attraction—which keep them herded together in the cloud—and gravity, which pulls them down. When the drops are small, their surface area is huge compared to volume; electrical (molecular) forces rule and the drops stay suspended in midair. Once the drops get big enough, however, gravity always wins.

Pint-size objects barely feel gravity—a force that only makes itself felt on large scales. The electrical forces that hold molecules together are trillions of times stronger. That's why even the slightest bit of electrical static in the air can make your hair stand on end.

These electrical forces would present major problems to flea-size Superman. For one thing, he'd have a hard time flying faster than a speeding bullet, because the air would be a thick soup of sticky molecules grasping him from all directions; it would be like swimming through molasses.

Flies have no problem walking on the ceiling because the molecular glue that holds their feet to the moldings is stronger than the puny weight pulling them down. The electrical pull of water, however, attracts the insects like magnets. As Haldane points out, the electrical attraction of water molecules makes going for a drink a dangerous endeavor for an insect. A bug leaning over a puddle to take a sip of water would be in the same position as a person leaning out over a cliff to pluck a berry off a bush.

Water is one of the stickiest substances around. A person coming out of the shower carries about a pound of extra weight, scarcely a burden. But a mouse coming out of the shower would have to lift its weight in water, according to Haldane. For a fly, water is as powerful as flypaper; once it gets wet, it's stuck for life. That's one reason, writes Haldane, that most insects have a long proboscis.

In fact, once you get down to bug size, almost everything is different. An ant-size person could never write a book: the keys to an ant-size typewriter would stick together; so would the pages of a manuscript. An ant couldn't build a fire because the smallest possible flame is larger than its body.

Shrinking down to atom size alters reality beyond recognition, opening doors into new and wholly unexpected vistas. Atom-size things do not behave like molecule-size things or human-size things. Atomic particles are ruled by the probabilistic laws of quantum mechanics. Physicists have to be very clever to lure these quantum mechanical attributes out in the open, because they simply don't exist on the scales of human instrumentation. We do not perceive that energy comes in precisely defined clumps or that clouds of electrons buzz around atoms in a permanent state of probabilistic uncertainty. These behaviors become perceptible macroscopically only in exotic situations—for example, superconductivity—a superordered state where pairs of loose electrons in a material line up like a row of Rockettes. With electrons moving in lockstep, electricity can flow through superconductors without resistance.

Scale up to molecule-size matter, and electrical forces take over; scale up further and gravity rules. As Philip and Phylis Morrison point out in the classic *Powers of Ten,* if you stick your hand in a sugar bowl, your fingers will emerge covered with tiny grains that stick to them due to electrical forces. However, if you stick your hand into a bowl of sugar cubes, you would be very surprised if a cube stuck to your fingers—unless you purposely set out to grasp one.

We know that gravity takes over in large-scale matters because everything in the universe larger than an asteroid is round or roundish—the result of gravity pulling matter in toward a common center. Everyday objects like houses and mountains come in every old shape, but mountains can only get so high before gravity pulls them down. They can get larger on Mars because gravity is less. Large things lose their rough edges in the fight against gravity. "No such thing as a teacup the diameter of Jupiter is possible in our world," say the Morrisons. As a teacup grew to Jupiter size, its handle and sides would be pulled into the center by the planet's huge gravity until it resembled a sphere.

Add more matter still, and the squeeze of gravity ignites nuclear fires; stars exist in a continual tug-of-war between gravitational collapse and the outward pressure of nuclear fire. Over time, gravity wins again. A giant star eventually collapses into a black hole. It doesn't matter whether the star had planets orbiting its periphery or what globs of gas and dust went into making the star in the first place. Gravity is very democratic. Anything can grow up to be a black hole.

Even time ticks faster in the universe of the small. Small animals move faster, metabolize food faster (and eat more); their hearts beat faster; their life spans are short. In his book *About Time,* Paul Davis raises the interesting question: Does the life of a mouse feel shorter to a mouse than our life feels to us?

Biologist Stephen Jay Gould has answered this question in the negative. "Small mammals tick fast, burn rapidly, and live for a short time; large mammals live long at a stately pace. Measured by their own internal clocks, mammals of different sizes tend to live for the same amount of time."

We all march to our own metronomes. Yet Davis suggests that all life shares the same beat because all life on Earth relies on chemical reactions—and chemical reactions take place in a sharply limited frame of time. In physicist Robert Forward's science fiction saga *Dragon's Egg,* creatures living on a neutron star are fueled by nuclear reactions; on their world, everything takes place millions of times faster. Many generations could be born and die before a minute passes on Earth.

And think how Earth would seem if we could slow our metabolism down. If our time ticked slowly enough, we could watch mountains grow and continental plates shift and come crashing together. The heavens would be bursting with supernovas, and comets would come smashing onto our shores with the regularity of shooting stars. Every day would be the Fourth of July.

An artist friend likes to imagine that if we could stand back far enough from Earth, but still see people, we would see enormous waves sweeping the globe every morning as people stood up from bed, and another huge wave of toothbrushing as people got ready to bed down for the night—one time zone after another, a tide of toothbrushing waxing and waning, following the shadow of the Sun across the land.

We miss a great deal because we perceive only things on our own scale. Exploring the invisible worlds beneath our skin can be a terrifying experience. I know because I tried it with a flexible microscope attached to a video camera on display at the Exploratorium in San

Francisco. The skin on your arm reveals a dizzy landscape of nicks, creases, folds, and dewy transparent hairs the size of redwood trees—all embedded with giant boulders of dirt. Whiskers and eyelashes are disgusting—mascara dripping off like mud on a dog's tail. It is rather overwhelming to look through your own skin at blood cells coursing through capillaries. It's like looking at yourself without clothes. We forget the extent to which our view of the world is airbrushed, that we see things through a shroud of size, a blissfully out-of-focus blur.

An even more powerful microscope would reveal all the creatures that live on your face, dangling from tiny hairs or hiding out in your eyelashes. Not to mention the billions that share your bed every night and nest in your dish towels. How many bacteria can stand on the pointy end of a pin? You don't want to know.*

We're so hung up on our own scale of life that we miss most of life's diversity, says Berkeley microbiologist Norman Pace. "Who's in the ocean? People think of whales and seals, but 90 percent of organisms in the ocean are less than two micrometers."

In their enchanting journey *Microcosmos*, microbiologist Lynn Margulis and Dorion Sagan point out the fallacy of thinking that large beings are somehow supreme. Billions of years before creatures composed of cells with nuclei (like ourselves) appeared on Earth, simple bacteria transformed the surface of the planet and invented many high-tech processes that humans are still trying to understand—including the transformation of sunlight into energy with close to a 100 percent efficiency (green plants do it all the time). Indeed, they point out that fully 10 percent of our body weight (minus the water) consists of bacteria—most of which we couldn't live without.

Zoom in smaller than life-size, and solid tables become airy expanses of space, with an occasional nut of an atomic nucleus lost in the center, surrounded by furious clouds of electrons. As you zoom

*For an eye-opening view, read *The Secret House*, by David Bodanis.

in, or out, the world looks simple, then complex, then simple again. Earth from far enough away would be a small blue dot; come in closer and you see weather patterns and ocean; closer still and humanity comes into view; closer still and it all fades away, and you're back inside the landscape of matter—mostly empty space.

So complexity, too, changes with scale. Is an egg complex? On the outside, it's a plain enough oval, like Jupiter's giant red spot. On the inside, it's white and yolk and blood vessels and DNA and squawking and pecking order and potential chocolate mousse or crème caramel.

The universe of the extremely small is so strange and rich that we can't begin to grasp it. No one said it better than Erwin Schrödinger himself:

> As our mental eye penetrates into smaller and smaller distances and shorter and shorter times, we find nature behaving so entirely differently from what we observe in visible and palpable bodies of our surroundings that no model shaped after our large-scale experiences can ever be "true." A complete satisfactory model of this type is not only practically inaccessible, but not even thinkable. Or, to be precise, we can, of course, think of it, but however we think it, it is wrong; not perhaps quite as meaningless as a "triangular circle," but more so than a "winged lion."

And as we shall see in the next section, the alchemy that can occur as we move from less to more or large to small is both unnerving in its unexpectedness and awesome in its explanatory power.

Chapter 6

EMERGING PROPERTIES: MORE IS DIFFERENT

Today, we cannot see whether Schrödinger's equation contains frogs, musical composers, or morality—or whether it does not.

—*physicist Richard Feynman*

The phrase "more is different" was coined (or so some say) by physicist Philip Anderson in response to what he felt was a misplaced emphasis—especially in his own field—on distilling everything in nature down to its fundamental elements. Certainly, everything on Earth ultimately boils down to protons, neutrons, electrons, light, and gravity. But what does that tell us about the properties of weather, chewing gum, or rain forests? Very little, Anderson and others believe. To get to the truth about something, you have to do a lot more than simply boil it down to its simplest parts.

To be sure, the physics of elementary particles leaped ahead with astonishing rapidity during the early and middle part of this century. First the electron was discovered, then the proton, then the neutron. In the 1970s, protons and neutrons were revealed to be tight sacks of even more fundamental particles called quarks. Electromagnetism and the so-called weak force responsible for radioactivity were shown to be closely connected. Any day now, it seemed, physicists should be able to describe the entirety of existence with a few simple formulas.

Theoretically, at least, it was even possible to use the wave equation formulated by Erwin Schrödinger to describe any atom—and therefore anything made of atoms—perfectly and completely. In reality, these equations are much too hard to solve (although some researchers using clever mathematics and fast computers have already designed new materials using them). Still, the idea that everything could be understood by reduction to its simplest components was too tempting for many people to resist.

The problem with this approach, for Anderson and others, was that you simply couldn't get there from here—from quarks to the cosmos. As Feynman noted, nothing we know about elementary particles and forces can tell us anything about green reptiles that croak in the night, or the music of Mozart, or the Ten Commandments.

"When and if we have found and understood the complete irreducible laws of physics," writes physicist Frank Wilczek, "we certainly shall not thereby know the mind of God (Hawking to the contrary). We will not even get much help in understanding the minds of slugs, which is about the current frontier of neuroscience."

The universe is full of things that cannot be understood—ever—simply by understanding smaller and more fundamental parts. Each time you go from quarks to atoms to chewing gum to life to galaxies, new things emerge that cannot be explained or predicted by goings-on at lower levels. "Psychology is not applied biology," Anderson wrote in his 1977 essay in *Science* magazine. "Nor is biology applied chemistry."

Mind is made of matter, but it is also clearly more. The most complete possible knowledge of water molecules will not allow you to predict a thunderstorm. Everything from clouds to life to superconductivity to music is the result of the profound qualitative shift that occurs when things get together in an organized way. Something comes into being on the scale of many that does not exist on the scale of few. More, in other words, really is different.

The physical world is full of examples. The arrangements of electrons in atoms ultimately produce such material qualities as color, electrical conductivity, texture, strength, and so forth. But a single atom cannot be rough or brittle or sweet. Much like the material phase shift that takes place when the temperature of water drops below thirty-two degrees Fahrenheit and freezes into ice, a change in quantity produces qualitative results. A single water molecule cannot freeze any more than a single carbon atom can have the hardness or sparkle of diamond.

The idea that more is different offers one way of resolving Schrödinger's cat paradox (as well as many other quantum mechanical conundrums). As Stewart and Cohen point out, a typical cat contains about 10^{26} atoms—10 with twenty-six zeros after it. The cat is an emergent property that simply doesn't exist on the single-atom scale. You can learn everything there is to know about the atoms that make up a cat, and that still will not tell you whether it will scratch the furniture or sleep on your head.

In the same way, individual people behave quite differently than crowds. One person cannot have mass hysteria any more than one illness is an epidemic. In fact, crowd behavior is much more predictable than the behavior of any individual. This fact applies equally to inanimate objects. Toss a coin, and you cannot predict whether it will come up heads or tails; no matter how many times you toss it, the probability of it coming up heads or tails remains fifty-fifty. However, if you toss a coin a million times, you can be certain it will come up tails roughly 500,000 times. While no gambling establishment can predict which number will come up on a single roll of dice, they can predict with some confidence the outcome of a great many rolls—that's how they make their profits.

In a sense, all the patterns of nature, from flowering trees to ocean swells, from mountains to koala bears, are the emergent properties of simple interactions between subatomic particles that over time add up to far more than the sum of their parts.

Time may well be the ultimate emergent property. A single particle can go backward or forward in space or time—it makes no difference—and there's no clear way to tell which is which. The only time that exists is the atom's own internal clock—the frequency at which it vibrates. But put a bunch of atoms together, and no one has any problem telling which way time flows: It's always in the direction of disorder. Left to their own devices, food rots, skin wrinkles, paint peels, mountains erode, stockings run. And yet, there is no hint of this large-scale headlong rush toward disorder within any single atom alone.

In one sense, "more is different" is the mathematical version of the old saying about the straw that breaks the camel's back. At some point, more changes everything.

There is an instructive analogue to "more is different" in social policy. First developed by Harvard economist Thomas Schelling to study racial segregation in housing, it has to do with how small quantitative changes in behavior lead to huge qualitative differences. Say a black family moves into a white neighborhood; then another, then another. At what point do white families begin to think about moving out (or vice versa)?

Schelling found that this tipping point, as he called it, is extremely sensitive—in that a very small increase (in this case, in the number of black families) can lead to a very large effect, enough to change a mixed neighborhood into a black neighborhood in short order.

Mathematician John Casti said that to his "physics-trained" eye, the tipping point looked like a sociological version of the kind of phase transition that takes place when water turns to ice. All it takes is a decrease of one degree, and a flowing liquid hardens into a solid that keeps its shape and stays put.

Such tipping points seem to play a role in many other situations where minorities are outnumbered. For example, some studies have suggested that women in physics classes feel isolated and unwelcome until the number of females in the class (or department) reaches something like 15 percent. At that point, the atmosphere tips enough to change physics from unbearably icy territory to a habitable (if not always warm) environment.

In other contexts, tipping points may account for seemingly unexplainable sudden changes in social behavior. New York City's unaccountable plummeting crime rate was described in an article by Malcolm Gladwell in the *New Yorker* magazine as a prime example of what the tipping point can do. The rate of violent crime—which had been among the highest in the world—suddenly fell so steeply that New York now ranked 136th among cities in the United States, about the same as Boise, Idaho. "There is probably no other place in the country where violent crime has declined so far, so fast," Gladwell writes, but how did it happen? For many years, the city had been working on improved methods of policing, graffiti removal, and the like, with few visible results. But nothing happened that could account for the huge scale of the change. There seemed to be an enormous discrepancy between these small increments of effort and a sudden huge payoff.

Gladwell compared the crime problem to an epidemic and used the arithmetic of epidemics to analyze the problem. Just as small changes seemed to precipitate a huge drop in crime, it took very little to turn a flu outbreak into an epidemic. All of a sudden a critical mass is reached when things accelerate seemingly out of control. Epidemiologists, too, he says, call this the "tipping point."

The lesson is that causes and effects do not have a simple relationship.* You can keep piling straws upon the camel's back, and nothing will happen until you put the critical straw in place. Then, at that critical point, the small change can have a huge effect. Before that critical threshold, however, even relatively large changes seem to have distressingly small effects.

This applies even when the effects are negative. For example, many studies have shown that women who drink alcohol during pregnancy may give birth to children with fetal alcohol syndrome. But evidence suggests that damage to the unborn child only begins after a certain threshold is reached—say, several drinks a day.

What this means for public policy, Gladwell says, is that we shouldn't jump to conclusions about the effectiveness or failure of social policies without taking the tipping point into account. We shouldn't conclude, say, that the welfare system doesn't help people get out of poverty because it hasn't accomplished that goal yet. We shouldn't conclude that money spent on inner city schools is wasted because it hasn't shown results comparable to money put in. It could be that we simply haven't yet reached the tipping point.

*See Chapter 7, "The Mathematics of Prediction."

Chapter 7

THE MATHEMATICS OF PREDICTION

It is impossible to trap modern physics into predicting anything with perfect determinism because it deals with probabilities from the outset.

—*Sir Arthur Stanley Eddington*

The *Galileo* spacecraft cruised the solar system for six long years before arriving at the giant planet Jupiter in December 1995—the final destination of a 2.3-billion-mile journey that looped twice around Earth and once around Venus.

A few months before its arrival date, *Galileo* unlatched a small probe that hitched a ride on its belly for the first six years of the trip. The seven-foot-tall cone-shaped probe was to drop through a hole in the Jovian clouds at precisely the moment that the mother ship passed overhead—allowing the probe to beam up the data it collected on this first ever descent inside a giant gas planet.

It was a very tricky maneuver. The probe's entry had to be as precise as a hypodermic needle slipping in beneath the skin. If the entry angle was too shallow, the probe would bounce off the planet's atmosphere like a stone skipping from the surface of a pond; too deep, and it would be destroyed before it could phone home any information.

As the world now knows, on December 7, at exactly 5:06 P.M. Pacific time, precisely as planned, the tiny probe dove through Jupiter's pastel-colored clouds cleanly enough to earn it an Olympic gold. Despite the duration and distance involved in the journey, its aim on arrival was picture-perfect.

This is the kind of spectacular success that leads people to believe science is good at predicting just about anything: the next earthquake, the next cancer victim, the next stock market crash, the global climate twenty or two hundred years from now.

Yet nothing was more frustrating during the 1994 earthquake in Los Angeles than watching the poor reporters trying to get the geologists to predict what was going to happen next—and the geologists' equal frustration trying to explain that prediction is not what they do.

The idea that science possesses a crystal ball that can look accurately into the future is as old as science itself. Pierre-Simon de Laplace, who served as mathematician to Napoléon, stated unequivocally that any intelligent being who knew the exact description of every particle for a given instant could accurately predict the future. "For such an intellect," he wrote, "nothing could be uncertain; and the future just like the past would be present before its eyes."

Prediction, in other words, was merely a matter of accumulating enough information. Theoretically, the future was already locked into the present by events in the past. All you needed to figure it out was knowledge.

Certainly, the astronomers have been able to predict the motions of the heavens with uncanny foresight. Eclipses of the Sun and Moon, conjunctions of the planets, the motions of the stars and constellations (not to mention artificial satellites such as *Galileo*) can be predicted hundreds and even hundreds of thousands of years in advance.

But prediction is neither the goal nor the forte of science. If truth be told, the physicists can't even perfectly predict where a Ping-Pong ball will bounce on the other side of the table. (The paradox that science can be at the same time very, very good and horrid at prediction was all too evident at a recent gathering to celebrate the last lunar eclipse until 2000. While the astronomers precisely pinpointed the exact moment when the shadow of Earth began to take a bite out of the Moon, thousands of celebrants at Griffith Park Observatory in Los Angeles missed out on most of the show. The problem was that clouds rolled in from the south and covered up the Moon. While scientists can predict eclipses centuries in advance, they can't predict the weather from moment to moment.)

A lot of the misunderstanding over the role of prediction in science can be traced to the seemingly innocent phrase "the theory predicts." Understandably, people interpret it to mean predict as in foretell the future. But it's really about predicting the present.

Einstein's theory of gravity, for example, predicts that light from a distant star should bend as it dips into the gravitational field of a massive object like our Sun. He wasn't predicting an event that might happen next week. Since time began, starlight has been getting bent out of shape as it skims by massive stars. Before Einstein's prediction, however, no one thought to look for this effect. Einstein figured out a way, and as soon as World War I had cooled down sufficiently for international scientific expeditions to resume, Sir Arthur Eddington traveled to the Cape of South Africa to test Einstein's prediction during an eclipse—when a star just beyond the Sun would be visible. Eddington found that the light from the star bent just the way Einstein said it would, confirming his prediction (and thus, adding weight to the theory itself).

Einstein's theory also predicted that space could bend so much it could pinch itself into black holes. Again, he didn't mean he thought black holes would suddenly appear at some time in the future. He meant they were there for the seeing, if only we could figure out where to look and how. Although evidence for black holes has not been nailed down as firmly as evidence for the bending of light, the general consensus is that telltale signatures of black holes, too, have been found—just as Einstein predicted.

The same story, with shifting details, is told again and again in science. When Heinrich Hertz heard about James Clerk Maxwell's theory that light was an undulation of electromagnetic energy in space, he "predicted" the existence of waves much longer than the eye could see. If visible light could be created by an ebbing and flowing of electric and magnetic fluctuations, what prevented much slower vibrations? Or much faster ones? The slow ones turned out to be radio; the fast ones, X rays.

Similarly chemists can "predict" the outcome of reactions based on their knowledge of the periodic table and quantum mechanics; they predict what will happen if certain elements are brought together under certain conditions, not what will happen next week.

Needless to say, there's a huge difference between this sort of prediction and forecasting events that haven't happened yet. A scientific prediction is less like a weather forecast than a train of thought. If *x*, then *y*. If clouds, then rain. If curved space, then black holes. The better the theory, the more accurately it will point to where and how you might look to confirm your suspicions.

But prediction is rarely the point. Predictions are used to test which theories are on the right track, going in the right direction. If the theory's predictions prove false, then it's clear the thinking has to change directions. Predictions are guideposts along the way to understanding, not goalposts.

Geologists, for example, study the structure of Earth and the planets, and the rocks and minerals that make them up. Every now and again, they come up with a theory that needs to be tested with a concrete prediction. For example, if continental plates shift around the surface, carrying chunks of land and seafloor with them, they ought to leave evidence behind of where they've wandered before (and they do). If earthquakes are caused by massive plates crashing against each other, then earthquakes ought to occur primarily along fault lines (they do).

None of this helps seismologists to predict at what time next week the windows in Los Angeles are going to shatter, the freeways buckle. But it certainly does help them understand how the earth moves. And that's what science is ultimately all about: how and why, not where and when.

This sometimes seems to be forgotten even by scientists—or perhaps more accurately, scientists as their ideas get presented in the popular press.* For the past decade or so, for example, a fierce debate has been raging over how the universe will end. Will it continue to expand outward, as it has since the big bang, for all of eternity? Or will it eventually reverse direction, bounce back, the gravitational attraction of all the matter pulling the universe back onto itself, compressing itself into an infinitely small space, then perhaps exploding again—a second big bang?

In the first case, the universe dies an ignominious death, disintegrating into disorder like a stale sandwich left out in the sun; in the latter case, we get a second chance at creation. What will the future bring?

The answer hinges on the amount and type of matter in the universe. Many cosmologists think that as much as 99 percent of all the matter in the universe is of an exotic variety that has not yet been detected—perhaps because it is undetectable. It seems to me that the question of why humans and the Earth we inhabit are constructed of a type of matter that is an oddity in the universe is far

*This writer not excluded.

more interesting than the question of how it will end. But either way, the accuracy of the prediction depends on the accuracy of knowledge about the present, and understanding is the ultimate point.

The kind of prediction science does so well might better be described as pattern perception.* *Galileo* got to its target not by predicting the future, but by following well-known patterns to their logical conclusions. Objects in motion follow well-understood paths as they coast and fall and loop around bodies in space. If you know the patterns, it becomes a matter of mere calculation to get *Galileo* to Jupiter on target.

In the same way, the existence of the planet Neptune was predicted by an astronomer who noticed a wobble in the orbit of Uranus. Something was pulling Uranus out of its expected pattern of motion; a deviation from the pattern pointed to the outside gravitational influence of another body. In 1846 Neptune was found to be orbiting just where predictions said it should be.

The fact that patterns repeat allows us to formulate laws of nature—really, recipes encoded in equations that describe relationships that repeat over and over again. Force equals mass times acceleration. The bigger they come, the harder they fall. Every action produces an equal and opposite reaction. It always works. The equation is a shorthand for a relationship with an enduring quality.

Many scientists accept the probability of life on other worlds for precisely this reason. Carbon atoms are exactly the same everywhere in the universe. Given the right conditions (pressure and temperature and so forth), they will join with other atoms into much the same patterns. Just as hydrogen and oxygen unite to form ice on Earth and do exactly the same on Jupiter's moon Europa.

Matter falls into formations guided by the forces that prevail in our universe, and carbon—like every other element—likes to re-

*That is, if you consider physical laws to be patterns of interactions encoded in equations.

arrange itself in certain predetermined ways. Under very high heat and pressure, carbon atoms join to form strong, tight tetrahedrons, bending and separating light into the sparkly colors of diamond. Arranged in flat six-atom hexagons like chicken wire, carbon is slippery graphite, good for pencils and lubricants.

Depending on its surroundings, carbon will tend to form the same kinds of patterns over and over again. A diamond is the same whether it is found deep inside Earth or in the rings of Saturn.

So it's not all that surprising that many scientists are willing to predict that carbon-based life could be rather common—again, given the right conditions, including water, moderate temperatures, stable environment, protection from radiation, energy. Such conditions may well have existed on ancient Mars—and may still exist in hidden watery worlds that may lie hidden underneath Europa's frozen crust.

Prediction that follows from pattern perception can be based on a deep understanding of nature, but it needn't be. After all, people could predict the motions of the planets before they understood the laws of motion. People do the same thing when they predict rain by joints that ache, or birds flying low. It doesn't take a scientist to predict the ebbing and flowing of tides, or a doctor to tell when a cranky baby might be getting sick. The periodic table of elements allowed Mendeleyev to predict the existence of various missing elements even though at the time, a sound theory of the atom did not exist.

Prediction, said physicist Frank Oppenheimer,* "is dependent only on the assumption that observed patterns will be repeated. The merchant, the politician, the parent, the artist, and the doctor all

*This entire chapter is very much informed by Oppenheimer's views on the subject. Founder of The Exploratorium in San Francisco, Frank was the younger brother of J. Robert, the "father" of the atomic bomb. His pacifist politics got him blackballed from physics during the McCarthy era.

depend for their success on the subtleties of pattern recognition. . . . The predictions of the physicist, the psychologist, or the economist in no way set them apart from the rest of humanity."

Sometimes, however, following patterns can lead even scientists grossly astray. That's because not everything follows a straight line from here to there. You could predict that if I start walking down my street and continue walking straight ahead at the same rate I will reach the corner store in half an hour. But if there is a roadblock between myself and the store—or an angry dog—the trip might take much longer.

More fundamentally, many if not most things don't change in straight-line patterns. For example, children do not continue to grow indefinitely; in late adolescence, they slow down. A built-in braking mechanism stops their growth before they get too tall for their bones to support them. Many other things swing in regular (or not so regular) cycles: the climate gets warmer and colder again; epidemics come and go; the stock market goes up and down.

In these cases, researchers trying to predict future trends often rely on a technique called curve fitting, in which a mathematical function is found that describes an existing pattern. The pattern is then extended into the future according to the same formula. Say you plot the rise and fall of the popularity of a particular movie star, or incidence of cancer, or interest rates. Then you find an equation that describes the curve. Then you continue the curve in the same direction it's going. In theory, a correct description of the dynamics creating the existing curve should allow you to extrapolate with confidence into the future.

The problem is, the same curve can often be described by very different equations. Author and physiologist Robert Root-Bernstein has described the dangers of overreliance on this technique of curve fitting as a way to extrapolate into the future, and he points to some notable failures—for example, in predictions about the spread of AIDS or the threat of global warming. Just twenty years ago, he points out, articles in science journals warned of a coming ice age and the potential for galloping glaciers. (The point isn't that global warming isn't real, but rather that predicting the future based on trends of the past can be tricky.)

He argues instead that we first need a deeper understanding of the basic phenomena involved, that we need to learn much more about climate and disease and population before we can sensibly begin to predict the future. Predictions based on extrapolation can't be any more accurate than the models we begin with. The *Galileo* spacecraft found its way to Jupiter right on target because celestial dynamics are well understood. The same cannot be said of many other sciences. In fact, the tipping point is as good an example as any of a growth curve that does not behave in anticipated ways.*

Alas, even perfect understanding doesn't always allow us to predict the future. Nature, it seems, has contrived to make even simple pattern perception unreliable under many common conditions. For example, no amount of understanding of the behavior patterns of atoms allows you to predict just where an atomic particle will be at a certain time. The best you can get is a probability. Or if you do pin down the particle's position precisely, you can't say anything very precise about its velocity. One can't be known without sacrificing the other.

Encoded into physics as the Heisenberg uncertainty principle, this weird fuzziness of the subatomic realm seems to put a natural limit on what we can know. Laplace's grand scheme for finding out

*See Chapter 6, "Emerging Properties: More Is Different."

the position and motion of every particle in the universe becomes a theoretical impossibility. The information simply cannot be dragged out of the atom and into the open. (Worse, physicist Stephen Hawking has proposed that information can get lost from the universe forever in the bottomless pits of evaporating black holes—leaving behind no trace whatsoever of its previous form. "One could imagine that particles and information could fall into these holes and get lost," he's written. "Maybe that is where all those odd socks went.")

Even without Heisenberg, however, prediction is an almost impossible proposition outside the physicist's make-believe world of ideal billiard balls. A simple prediction such as where a particular drop of water will fall as it crashes over Niagara Falls is beyond our capabilities. It simply requires too much information.

Or consider the information required to predict the trajectory of a ball batted into right field by a baseball player. For openers, you would need precise information about the velocity and spin of the ball, the elasticity of the materials, the interaction of surfaces, the weight and structure of the batter's hands, not to mention winds, temperature, humidity, and so forth. Newton's laws are simple enough, but the ingredients added by reality make the problem overwhelmingly complex. Plus, someone could throw a popcorn box at the ball, or a bird could fly into its path.

"The ballplayer will predict such events more reliably than the physicist," writes Oppenheimer. "Yet this situation is ludicrously simple, and one which is described by one of the most highly perfect mathematical ideas in physics."

Taken together, the Heisenberg limits on measurement and the general complexity of real life make accurate prediction of certain kinds of events close to impossible. "This rise of statistical prediction, of probability," writes physicist Philip Morrison, "is perhaps the most characteristic of all the developments of twentieth-century science. It represents the realization that we cannot claim to know all the causes of things, for those causes are far too numerous."

So perhaps it is fitting that one of the hottest fields in mathematics today is known as "complexity" theory—in essence, a theory of unpredictability. In a way, it's the modern version of the Heisenberg uncertainty principle.

It's a strange kind of field because it covers everything from economic systems to human consciousness, from the formation of galaxies to the behavior of clouds, from the study of Earth's core to the evolution of stars. But the central idea is simple: Take any simple thing—a water droplet, a star, the firing of a single neuron. Each, on its own, might be perfectly predictable. But put a bunch of them together, and you've got clouds, galaxies, mind—all unpredictable phenomena.

Take a pendulum. It is the epitome of predictability—so predictable we base our clocks on it, assured of the knowledge that every tick and tock will repeat in a numbingly predictable way. Yet if you put several pendulums together so that the motion of one affects the motion of another, they begin to twitch unpredictably, veering and swooping like eddies in a fast-running stream.

The reason is that in complex systems, every part influences what all the other parts do, creating a tight weave of causes and consequences much too knotted to untangle. It's an avalanche of influences where every pebble tugs on every other, causing sudden and unforeseeable effects.

Unpredictable systems like these were until fairly recently outside the boundaries of rigorous physics, where well-behaved billiard balls were the study objects of choice. The laws of nature did not seem to extend to such unruly subjects as weather and the turbulence of water flow.

To be sure, each player in these systems behaves according to well-understood Newtonian laws. Every atom of oxygen and hydrogen, every molecule of water, every gust of wind, follows the same rules as the stately procession of planets around the Sun. But one system is predictable; the other is not.*

Complexity expert James Crutchfield explains the difference, in part, as the result of so many moving parts. The "web of causal influences among the subunits can become so tangled that the resulting pattern of behavior becomes quite random," he says. Only in isolation can most systems remain beyond the reach of chaos.

In complex systems, feedback loops connect the parts in such ways that one part affects the next, which in turn affects all the others, and so on. Climatologists struggling to understand planetary weather systems are stuck in the middle of just such an endless feedback loop. The carbon dioxide exhaled into our air by cars and factories holds on to heat and could lead to global warming. But green plants breathe in CO_2 to live and may, just possibly, thrive in a high CO_2 environment. If global warming leads to a proliferation of plants, then it could result in global cooling as plants inhale most of the CO_2. The role of clouds in climate control is even more tangled and ambiguous.

The same tangle obscures the exact origin of many health problems, for example, why one person gets cancer and another doesn't. Genetics, environment, and behavior all interact in ways too complicated—at least for now—to sort out (with a few dramatic exceptions, like the link between smoking and lung cancer). That's why when someone pronounced that the transformation of the man now accused of being the Unabomber from a quiet mathematician into a killer (the Unabomber) could be explained purely by genetics, most observers were skeptical. The influences that shape people are not that simple.

*In the long run, even Newton's laws can lead to chaos, due to the unavoidable irregularities that creep into even the most predictable systems, then get multiplied by exponential amplification.

The feedback loops, in turn, are powered by the process of exponential amplification—where each small change quickly multiplies into a major consequence. Even a game of billiards, Crutchfield points out, would quickly become unpredictable in the real world. In his simplified scenario, a player makes a shot, which sends the balls colliding with all the other balls. How long could a player predict the outcome, he asks. "If the player ignored an effect even as minuscule as the gravitational attraction of an electron at the edge of the galaxy, the prediction would become wrong after one minute."

Ivar Ekeland, in *The Broken Dice,* calls this process "exponential instability," and it's behind the well-known fickleness of weather. "We know, for example, that in meteorology the magnitude of a disturbance doubles every three days if nothing interferes with its development," he writes. If the conditions vary slightly—for example, "if someone lights a candle"—it could have no consequence at all. On the other hand, he writes, "the effect may amplify over time at an exponential rate; if it doubles every three days, it will be multiplied by 1,000 every month, 1,000,000,000 every two months, and 10^{36} every year."

What this means, he says, is "if we want to know what the weather will be like one year from today, we have to take everything into account, from the butterflies flying in the Amazon jungles to the candles burning in churches."

Complexity and chaos theory are all the rage these days*; but it's interesting to note that more than 30 years ago, well before these ideas became fashionable, Frank Oppenheimer wrote about the perils of predicting something as complex as weather:

> It is far too difficult to gather even one piece of data—to conduct even a single experiment. It would be hard enough, for example, to try to predict changes in humidity without becoming confused by extraneous

*While these subjects are fashionable, some scientists point out that they have not proved particularly useful and have produced few predictions that can be tested.

> effects or changing the outcome through the intrusion of our own probes. But add in all the other factors that influence the weather, and the task becomes impossible. In the first place, there are too many things that we do not understand about nature. In the second place, too many things can happen to make our predictions go wrong.

Curiously, some of the most accurate predictions are based on chance—the odds of something happening. In a sense, this explains why prediction works so well in some corners of physics, despite all the inherent uncertainty. Even though predictions of how, say, radioactive atoms decay or how particles interact with each other are strictly probabilistic, they are also very precise. That's because, on average, probability becomes highly predictable. So even though we can't say who will be killed in a car accident next year, we can easily predict how many people will be killed.

In some cases, we are forced to rely on prediction to extrapolate from the present into the future because there is no other way to get there. We cannot wait around until the end of the universe to see how things will turn out or travel inside a star to learn about nuclear reactions. We can't wait for probable trends like global warming to run out of control before we take steps to tame them. Despite all the caveats, prediction serves a critical—if limited—role in science. Sometimes prediction is all we have.

When Frank Oppenheimer was working at Los Alamos during the construction of the first atomic bomb, he also acted as a kind of "safety inspector." (His opposition to the use of the bomb and his call for worldwide ownership of atomic power got him thrown out of physics during the post–World War II McCarthy era.) Here he remembers one particular critical prediction:

> I can remember, before the test of the first atomic bomb, making a calculation which indicated that the heat produced by the released radioactivity would force the radioactive cloud, even amply mixed with cool air, to rise through any atmospheric inversion. The radioactivity would thus,

> according to my calculation, not spread out into a smog at low levels of the atmosphere. My conclusion proved to be correct: the mushroom has always had a long stem. But the fact that I had reached this conclusion did not deter me from helping to map an escape route through the tangled roads of the desert to the south of the test site. Nor did the calculation dispel the terror that the evilly glowing radioactive cloud instilled as it seemed momentarily to hang over our heads after the explosion. Furthermore, this calculation told nothing about the details of subsequent radioactive fallout.

Physicists, Oppenheimer believed, are generally well-informed about the uses and hazards of prediction. He was much more worried about the ways scientific notions of prediction were being incorporated into the social sciences. For example, any prediction involving people tends to be misleading, precisely because the numbers needed to make accurate predictions are so huge. The result is that the predictions become meaningless—because they don't apply to any individual person.*

Indeed, the emphasis on prediction in science, he argues, has probably contributed to the public's mistrust of science. "Not wanting their behavior controlled or predicted, people react by rejecting both the social and physical sciences. . . . If human sciences are to benefit society, the level of understanding must become more important than the clarity of the crystal ball."

*See Chapter 5, "A Matter of Scale."

Chapter 8

THE SIGNAL IN THE HAYSTACK

Most of the time is spent testing for artifacts. That's the pain. To make sure it's not you or car exhaust from the street, or contamination from the air, or something it picked up in Antarctica.

—Stanford chemist Richard Zare, on experiments to detect organic compounds on the now-famous meteorite from Mars

In 1996 Stanford and NASA scientists announced that they had found evidence of ancient life on Mars. They based their conclusions on markings found on a rock they believe was kicked off Mars by a meteor that crashed into the red planet millions of years ago, flinging tons of rock and dust into space. This particular rock, they believe, orbited the Sun as a chunk of space debris for about 16 million years before landing on a blue ice field in the South Pole, where a sharp-eyed Antarctic researcher out for a joyride on her snowmobile spotted it, picked it up, and brought it home.

How do they know that the curious tube-shaped forms that landed in their lab were fossils of ancient life, rather than dried-up mud cracks, as one researcher put it? The simple answer is, they don't—at least, not yet. But they're working on it (along with dozens of other teams of researchers who would love nothing better than to prove them wrong).

The more complicated but truer answer is that facts rarely present themselves cleaned up and alone, ready to be admired and fussed over. Instead, nature bestows her blessings buried in mountains of garbage, and scientists rarely know what they have their hands on until they've sifted through the mess, laboriously, patiently, piece by piece.

The astronomers who found the first planets around stars other than our Sun faced similar problems. The announcements over the past few years about discoveries of new planets did not mean anyone had actually sighted a planet; they meant that astronomers had seen some unusual wiggles in the position of a star that indicated the star was being pulled off course by some unseen companion. When they saw the wiggle, they did not say: "Eureka! A planet!" Instead, they probably said something closer to: "Oops. Something must be wrong with our experiment."

And so it goes with almost any important discovery you can name. A few years back, an astronomer announced that he had seen the face of God—or the first wrinkles in space-time traveling to us from the creation of the universe. Another group of researchers reported they had found the so-called top quark—the last known member of the family of fundamental particles to be pinned down.

The data these scientists were looking at hardly spelled out a clear signature: Most of the time, they were looking at strings of numbers (what else is there, you might ask?). From a bunch of digital signals, they read patterns like tea leaves, telling us that there's more dark matter in the universe than we thought, or that a certain gene sits on a certain place on a chromosome, or that an asteroid is hurtling toward Earth. In fact, what they really "see" is an immense amount of static noise in which is buried a signal. Maybe.

Filtering out the static is a process central to both science and human perception (which are partners, after all, in much the same enterprise). You can't perceive anything if you can't block most of the information that comes your way. You can't hear the voice on the phone if other people are shouting in the background; you couldn't see anything at all if your iris didn't shut out most of the light—allowing only a tiny bit to sneak in through the pupil.

Scientists, in turn, have to shield their experiments from extraneous influences. Slices of the Mars rock are examined in a vacuum so that no earthly critters creep in and muddy the results. Particle physicists bury their detectors underground and cover them with tons of shielding so that cosmic rays don't make stray tracks that could be mistaken for new kinds of particles.

The astronomers may well have it worst of all. For one thing, noise drowns out most of their potential observing time. At night the star-spangled sky glitters like a black velvet showcase for multicolored diamonds—at least on a clear night far away from smog and city lights. During the day, however, there's absolutely no sign of those glittering prizes.

Where do the stars spend their days? They're up there, of course, decorating the sky as always, but you can't see them in the glare of the light from the Sun. You can't see the stars in the day for the same reason you can't hear a whisper in a noisy restaurant—the insistent Sun shouts them out.

Even at night, there's a lot of noise in the sky that can make it hard for astronomers to see the stars; there's light from the Moon and the city; there's heat from the telescope itself (many are refrigerated); there's wind that stirs up the images and makes them as blurry as a penny on the bottom of a pool. A lot of the art in astronomy (and other sciences) is figuring out how to get around the noise without losing the signal.

To some extent, the noise problem is merely an artifact of choice and circumstance. Consider the pile of dust and dirt accumulating under the refrigerator. The crumbs were recently part of the cake you had for breakfast; the cat hair was part of the animal's fur; the leaf belonged to a tree; and the paper clip found its way to the floor from something you opened in the mail. You don't consider any of these things as candidates for the trash heap until they wind up in the "wrong" place.

There is an old riddle that vividly demonstrates just how noise can interfere with thinking, even when that noise is information. Imagine you are a bus driver. At the first stop of the day, nine passengers get on your bus. At the second stop, two people get off. At the third stop, four people get off, but three new people get on. What color are the bus driver's eyes?*

Noise, in other words, is whatever you don't want to be where it is—whether it's the conversation in the background or the weeds in the garden. It's what you need to get rid of to see what we want to see, to learn what we need to learn.

These things we dismiss as noise often have a great deal to tell us, however. Both scientists and artists learn to pay attention to the crumbs that other people are about to sweep under the rug. They learn to be good noticers. The same could be said about good teachers, good parents, effective politicians.

And certainly inventors. Saccharin was discovered more than a hundred years ago when a scientist doing some chemistry experiments stopped during his research to have dinner. He noticed that his dinner tasted unnaturally sweet and that he had a strange white powder on his hands. The powder made his food—and also his fingers—taste sweet. He paid attention and gave the world a no-calorie substance hundreds of times sweeter than sugar.

*Since you are the bus driver, the driver's eyes are the same color as your own. The rest of the information—in this context—is noise.

Post-it Notes were invented in much the same way. A chemist investigating ways to make a better glue found instead a glue so ineffective it wouldn't permanently attach to anything. Instead of throwing the invention in the garbage, the chemist saw that it would work perfectly in a different context—sticky notepaper that could be peeled off without leaving a trace.

In fact, any signal takes its meaning from context; in a different context, the same message can have no meaning at all. If you send someone a message in code, but they have no way to decode it, your message has no more information than total nonsense. A tree falling in the forest may make a sound even if no one's around to hear it, but it doesn't convey a signal. Conversely, a scene devoid of signals to one person may contain a wealth of information to others. Think of Sherlock Holmes.

In *The Collapse of Chaos,* Stewart and Cohen point out that no signal has inherent meaning—outside the context of whatever hears it, or sees it, or decodes it. A compact disc, for example, may contain all the information necessary to play a piece of music. But without a CD player, it is only a pretty silvery disk—suitable for playing Frisbee, perhaps, but not much more. In the same way, a strand of DNA contains no message in the absence of molecules that can read the genetic code. "The number 911 has no inherent meaning," they write. "In the context of the U.S. telephone system, it means 'emergency'; in the context of a lottery it may mean 'you lose'; and in the context of housing it means that you live on a fairly long street."

Even DNA can transmit a different signal in different contexts, thereby producing very different results. "The caterpillar has the same DNA as the butterfly," the authors remind us, "the maggot has the same DNA as the fly, the human embryo the same DNA as the grandmother she eventually becomes. . . ."

As scientists know all too well, it's easy to mistake a signal for noise, and vice versa. It happens all the time. Yesterday's newfound planet is exposed as a glitch in the instrument that took its picture. That newly discovered particle turns out to be a cosmic ray. It happens the other way, too—when the background noise turns out to be a new particle, or the wiggle that was dismissed as an aberration turns out to be a planet.

Harvard science historian Gerald Holton poses the question this way:

> How to determine which of all possible demonstrable events are indications of scientifically usable phenomena; which of them are really connected to the fixed regularities of nature, and which are merely passing phantoms, clouds with ever-changing form never twice the same, and thus reflecting only ephemeral concatenations? We might call it the problem of telling the difference between signal and noise.

One night recently, I joined astronomers Sallie Baliunas and Chris Shelton on top of Mount Wilson in the San Gabriel Mountains for a night of observing. The hundred-inch Hooker telescope, which dominates the site, is practically a shrine in astronomy. It is the place where Edwin Hubble first saw that our Milky Way galaxy was not alone in the universe, but in the company of billions of similar "island universes." The Hooker also saw the telltale stretching of starlight that reveals our universe is still expanding from a central point in space and time, the explosive origins of everything—the so-called big bang.

During the 1980s the Hooker was mothballed. It stood as a historical curiosity, a World War I–era telescope with a mirror fashioned from French wine-bottle glass at the same factory that made the mirrors for the Hall of Versailles for Louis XIV. In 1995, however, it got a new set of optics that give it one of the sharpest views in the Northern Hemisphere. And, surprisingly, the clear air makes it one of the best observing spots in the world. (Ironically, the same inversion layer that traps the smog and suffocates downtown Los Angeles allows Mount Wilson to rise above it all for an exceptionally clear view.)

Harvard's Baliunas is using the Hooker (among other things) to look for Earthlike planets around other stars. The night I was there, she and Shelton—who built the optics—were taking the system out for a test run. And first on their list to look at was a special star that they wanted to examine because it played a central role in a possibly important discovery. Baliunas had learned that morning that another astronomer thought the star might harbor an unseen planet.

The problem was, a star with a planet orbiting it would send almost exactly the same signal as a star that had a small companion—that is, a double-star system. Planets are seen by the way they tug on the stars they orbit, moving the stars slightly from their normal paths. In other words, the signal that the astronomer who found the planet saw could have actually denoted a double star. Double stars are quite common, and therefore not anywhere near as important as planets, which are rare (at least as far as we know). If the star with the planet was actually a double star, then the planet would be an illusion. And only the Hooker could focus the star sharply enough to find out.

Shelton burped the telescope to get out any ambient air, then homed in on the star. To everyone's delight and astonishment, it appeared to be a double!—two stars piled up on top of each other like traffic signals. That meant no planet had been discovered. The signal meant a double star: bad news for the astronomer, but a good night for Baliunas and Shelton.

To make sure, however, they had to look at another star they knew was a single—to rule out any influence of the optical system itself. "You don't know anything until you compare and contrast," says Shelton. If the other star looked double, too, then it would mean that "noise" had crept into the system. If the other star looked like a dot, or a single, then it meant that the first star really was a double, and they'd made an important discovery.

To make a long story short, there was a lot of exuberance in the dome that night; the second star looked like a dot, and there were cheers and high fives all around. Then, just as suddenly, the dot cloned itself into two. The optical system was shaky. They had not made a discovery after all—except the all-important discovery that the system wasn't working quite right.

Particle physicist Leon Lederman had a similar experience with a particle called the upsilon that soon became known as the "Oops, Leon." Again, it was noise masquerading as signal. Later, the upsilon was pinned down with a better set of data.*

The discovery of antimatter is a good example of the opposite process—finding a real signal in what appears to be noise. A theorist discovered antimatter as a minus sign in an equation; an experimentalist saw it as a cosmic ray track that curved the wrong way. What's remarkable is that neither gave in to the temptation to dismiss such an unexpected discovery as a mere aberration.

Complicating the noise issue is the fact that one person's noise is another person's signal. I was struck by this on a visit to Kitt Peak in Arizona, where the National Science Foundation sponsors some state-of-the-art telescopes. As we went from one to another, it became clear that the signals one astronomer was seeking were considered so much noise in the eyes of another. At the solar telescope, for example, a group of researchers studying Earth's atmosphere was trying to get rid of telltale lines in the Sun's spectrum that indicated the presence of various elements on the Sun's surface. Since they were interested in elements on Earth, they had to get rid of "noise" from the Sun. Astronomers studying the Sun, however, have the opposite problem: for them, signals coming from Earth's atmosphere interfere with what they wish to find out. Or as Kitt Peak astronomer Richard Green said to me: "One person's floor is another person's ceiling."

---

*See Probable Truth in Chapter 13, "The Burden of Proof."

Luckily, there are a whole range of tools to take care of the noise problem, some of them mathematical and scientific, some of them built into the human perceptual system. In fact, the ability to shut out noise can be an important sign that a newborn is developing properly. Years ago I had the pleasure of watching esteemed Harvard pediatrician Berry Brazelton going through his standard exam of a very young baby. He shone a light in the baby's eye. The baby winced. He shone the light again, and the baby didn't respond at all. Next, he rang a bell in the baby's ear. The baby startled. He did it again. This time, no response.

All this indicated that the baby was developing normally, he said—and learning how to shut out "noise" is a major part of that effort. In everyday life, our brains continually ignore all kinds of similar stimuli: the nose that's always protruding into your field of view; the feeling of clothes on your back or rings on your fingers; the sounds of the refrigerator or the air-conditioning system.

Another way to get rid of noise is to wear blinders—literally, or figuratively. When you want to see clearly over the glare of streets or water, you don Polaroid sunglasses, which selectively filter out only those light waves coming at you from a horizontal plane (that is, glare). (You can play signal-noise games with Polaroid glasses by rotating them ninety degrees; now the horizontal surfaces such as streets will light up, but the vertical surfaces such as windows go dark.)

Some detectors designed to track down elusive subatomic particles are almost *all* filters. Elaborate triggering systems operated by computers automatically toss out more than 99 percent of the data retrieved, because particle accelerators simply create too much stuff for anyone to look at. The triggering systems screen every "event" for noise and throw most of them away.

To avoid throwing out the baby with the bathwater, however, researchers need to know everything there is to know about the types of noise likely to mess up their experiments. This means they have to become experts in the irrelevant, connoisseurs of distraction.

Consider the plight of the team of researchers who finally saw the first wrinkles in space-time coming at us from 10 (or so) billion years ago, the fossil imprints of the conditions at the origins of the universe.

The fact that seeing such a thing was even possible grew out of a famous signal-noise confusion. At about the same time that theoretical physicists had figured out that our universe could have been created in a violent explosion, various astronomers had been puzzled by an unexpected hiss in the sky. Thinking it was noise, the researchers figured it was the result of a creaky instrument or some other interfering source—perhaps bird droppings in the antenna.

The theorists, meanwhile, had figured out that if the universe really did start with a big bang, it should be possible to detect traces of leftover radiation from the explosion; it should still pervade the sky, coming at us from the origin, now vastly stretched and cooled over the course of its 10-billion-year history.

The punch line, of course, is that the "noise" bothering the astronomers turned out to be that very leftover radiation from the big bang. But that was only the beginning. As soon as people realized that they were reading messages from the origins of the universe, they set about trying to decode them—in particular, to find the seeds of structure in the universe, the tiny lumps in space-time that eventually grew into the great strings and clusters of galaxies that drape the darkness like garlands. The structure had to come from somewhere, so researchers set out to see if they could perceive it in the leftover radiation.

Berkeley astronomer George Smoot, who led the effort, describes the task this way: "We were looking for tiny variations . . . something less than one part in a hundred thousand—that is something like trying to spot a dust mote lying on a vast, smooth surface like a skating rink. And, just like a skating rink, there would be many irregularities on the surface that had nothing to do with those we sought."

Their satellite, the *Cosmic Background Explorer,* or *COBE,* was able to pick up the signals, but it also picked up enormous amounts of static noise. This noise could come from stray heat sources or magnetic radiation or artifacts in the software analysis or a host of other things.

"It is difficult to convey how obsessed we were with trying to eliminate these errors," writes Smoot. "I had started writing a list of potential things that could fool us back in 1974. Ever since, I had continually updated the list, adding new candidates. . . ."

A new and better *COBE* is already in the works. When the astronomers go back for a second look, they'll have to be on guard against a vast array of distractions. Foremost among them is the not so obvious fact that in order to get a good look at the cosmic background, you need to get rid of everything in the foreground. That includes signals that come from the instruments, from the Earth and Moon, from our own galaxy and other galaxies; it includes signal-absorbing dust inside our galaxy and beyond it. It includes the motions of objects as they sweep past each other in small eddies and large currents.

"There are many sources of confusion," says UCLA astronomer Ned Wright, who also worked on the project—the motion of Earth through the solar system, the motion of the solar system through the galaxy, the motion of the galaxy through the local cluster. "We don't know the orbital velocity of the solar system through the Milky Way," he says. "We don't know the velocity of the Milky Way through the local group."

In effect, these cosmic background explorers are trying to pick up a whisper in the roar of a crowd. Particle physicists face much the same problem. "It's not clear that they know how to handle the backgrounds," said Fermilab's Lederman, who won a Nobel Prize for the discovery of a particle so elusive it has been called (quite accurately) a spinning nothing. "When you're dealing with complicated backgrounds, it's not a numerical question. It's a visceral question. How do you feel about these backgrounds? How well do you understand these backgrounds?"

There are other ways to deal with unwanted information, to zoom in on what you really want to know. Microscopes do this by magnification, effectively pushing all extraneous matter out of the field of view. Telescopes zero in on certain targets to the exclusion of others. For every decision to focus on one thing, a piece of context is lost. These trade-offs are inevitable. In a way, it's like getting lost in thought, so concentrated on some delicious idea that the rest of the world simply fades away.

Astronomer Vera Rubin saw something no one else had ever seen recently because she spent several years looking in-depth at a galaxy that others had only skimmed. What she found was completely unexpected: stars rotating in opposite directions within the same galaxy. She compared her way of looking at galaxies with Georgia O'Keeffe's description of looking at flowers: "Nobody sees a flower," said O'Keeffe. "We haven't time, and to see takes time, like to have a friend takes time."

Astronomers who need to map the locations of millions of galaxies to make global sky surveys could never pick out Rubin's unusual specimen. "If you look at one million galaxies, you're going to miss the freaks," says astronomer Richard Green.

Still another noise-elimination tactic is to make the signal itself louder and clearer, so it can stand out from the fray. Astronomers and particle physicists use photo multipliers to do this, politicians use megaphones, the rest of us wear hearing aids or eyeglasses. The Hooker telescope's new optics act like a set of powerful spectacles that focus the light from a star so brightly that it stands out clearly even above the noisy skies of Los Angeles.

Smoot knew that when the signals came from the cosmic

background, they wouldn't be obvious or leap out, but slowly become apparent as they gained more and more strength. The weak signal, he explains, would get stronger by repetition, "like the ever-darkening mark left by repeatedly rubbing a pencil lightly over a piece of paper."

Another noise-reduction tactic is to make a very restricted, focused search for exactly what you are looking for—the complement of wearing blinders. When you tune into one radio station, you also tune the others out. *COBE* went actively hunting for a particular kind of signal. Our eyes and ears do much the same thing as we choose to actively listen or look at one thing and ignore the rest. "Sense organs do not passively accept incoming data; they go fishing for it," Cohen and Stewart point out—noting that more neural connections run from the brain to the ear than the other way around, and 10 percent of optic nerve fibers go the "wrong" way.

There are also many purely mathematical ways to get rid of noise and focus on signals. For example, one can assume that anything random is "noise" and eliminate it. Of course, that means you need to know what "random" is. Another way is to pick up only what changes, ignoring the constants in the same way as your brain ignores the motion created by the turning of your head or the image of your nose in your field of view. Computers can do this rather easily.

One very "hot" development in this field is wavelet analysis. Princeton's Ingrid Daubechies—one of the leaders in this approach to signal processing—calls wavelets "mathematical microscopes." They're squiggly shapes that can be superimposed on confusing signals, working like zoom lenses to zero in on just the interesting part of a picture, for example, while keeping the overall landscape in view.

Essentially, wavelets are atoms of information, and they come in a vocabulary of types, like the table of elements. Each wavelet represents a slightly different shape. From this basic vocabulary, you can build just about anything—just as you can build anything in the universe from the hundred-odd elements. Also, just as you can deconstruct anything in the universe into a hundred-odd elements, so you can deconstruct (almost) any signal into wavelets.

The Navy has found wavelet analysis useful for finding enemy submarines in the noisy environment of the ocean; astronomers used them to pick out a subcluster of galaxies in a galactic group; the FBI is using them to store and retrieve fingerprints more efficiently; electronics engineers are using them to create TV signals that work both on the new high-definition TVs and older sets.

One major benefit of wavelet analysis is that wavelets allow you to look at either the big picture or the local picture without losing information about the other. (A little like watching a sailboat race with binoculars, zooming in to see an individual sail number, but not losing your view of the other boats in the race.)

Even noise itself can be put to work enhancing signals. Called "stochastic resonance," this method makes use of the fact that sometimes random background fluctuations can boost weak signals to the point where they can be heard. An extremely rough analogy is a marble, say, sitting in one of those "cups" in an egg crate. Say you get a "signal" only when the marble hops from one cup to another. But in a quiet environment, the marble doesn't have quite enough energy to make that leap. If you put the egg crate in your car and drive over a rough road, however, you may find it much easier to get the marble to jump into another cup. The random shaking of the ride is enough to help boost the marble over the barrier—making your "signal" visible.

All of these signal-processing techniques are used to sharpen otherwise ambiguous signals in a wide variety of settings, from CAT scans of the human body, to sorting out social or political trends, to finding new planets. And all of them in the end remind us of the power of invariants—the things that do not change, no matter what. Only by getting rid of the noise, the distractions of intruding signals or warped frames of reference,* distorting effects of scale or the artifacts of measurement, can we come to know what the natural world is really like.

When Shelton and Baliunas wanted to check whether their double star was really a double star or simply a glitch in the system, they moved to a different star to see if things would change. When Smoot was seeking to verify that he'd really found the primordial wrinkles in space-time, one of the most convincing clues was the fact that the wiggles he saw were scale invariant—meaning that each patch of sky revealed the same spectrum of wrinkles, from smallest to largest.

That's exactly how the search for truth is supposed to work. You see something and then you try everything you can think of to make it go away; you turn it upside down and inside out, and push on it from every possible angle. If it's still there, maybe you've got something.

"You have to kick the measurement every way you can to see what shakes loose," says Baliunas—in a fitting metaphor for an astronomer who likes to soup up jalopies in her spare time. In other words, you do everything possible to eliminate all sources of possible error and confusion that could obscure what you're trying to see. Then you're ready to look for the truths that can be found in the mathematical art of pattern perception.

*See Chapter 14, "Emmy and Albert."

PART III

Interpreting the Social World

> To me, fair play means that the rules encourage everyone to play.
> They should reward those who win,
> but they must be acceptable to those who lose.
>
> —*Lani Guinier*

People who live in a democracy spend a lot of time making decisions about what's fair. It's built into our system of government that we should try to give everyone an even chance. That means we have to find ways to divide up everything from political power to property, from wilderness to air time, in what we consider to be fair ways.

Most of the time, we make decisions by voting. Voting is supposed to be fair because everyone gets a say in the outcome. If they don't get what they want, well, then they have to work harder to convince the rest of the people of the value of their particular point of view.

But there are many other ways to achieve fairness—for example, "taking turns," where various players each get a chance to choose. Or dividing things up, where one side gets a piece of the total pie (or whatever) that is roughly similar to the pieces that other sides get.

We use these methods to determine everything from who gets to be president to who gets to be prom queen, from the size of child-support payments to the size of the national budget, from the subjects taught in school to the way a family spends its summer vacation.

As might be expected, mathematicians have a lot to say about fairness—and as also might be expected, what they have to say is often surprising. The bad news is that some of our most cherished systems for dividing things up are grossly unfair; the good news is that in recent years, mathematicians have come up with a bevy of useful (if partial) solutions.

Chapter 9

VOTING: LANI GUINIER WAS RIGHT

What looks like a perfectly sensible, fair-minded set of rules for dividing up a resource can lead to a result that flies straight in the face of intuition and common sense.

—*John Casti, in* Complexification

In 1993, newly elected President of the United States Bill Clinton nominated a little-known law professor named Lani Guinier as his candidate to head the Justice Department's civil rights division. This soft-spoken legal scholar was known in civil rights circles for her innovative, if complex, views on voting rights. Her papers were well received by academic colleagues, who unanimously recommended that she be granted tenure at the University of Pennsylvania Law School.

Odd it was, then, that within weeks of her nomination, Guinier became the target of attacks so pervasive and persuasive that her old friend the president dumped her. She was called a "radical," "a madwoman," "a quota queen," a "racist," and even "loony Lani."

However, at least one group of professionals was squarely in her corner—although admittedly, not a group with enormous political clout. It was a well-regarded group of mathematicians. Guinier's critics, they pointed out, had displayed surprising mathematical illiteracy.*

Of course, ignorance does not explain the heartfelt rage directed at Guinier. The law professor's writings seemed to be threatening in a very personal way. People felt they directly attacked the nation's democratic ideals. Or as Clinton himself put it, he could no longer support her nomination because, he huffed, she "seemed to be arguing for principles of proportional representation and minority veto."

In a nutshell, Guinier was accused of questioning the very core of American democracy—the sacred ideal of majority rule. She was telling people what they didn't want to hear—that our election systems were neither fair nor democratic. What they produced was not democracy, but a tyranny of the majority. That phrase—"tyranny of the majority" (which is also the title of Guinier's subsequent book)—was lifted from the lips of none other than founding father James Madison. It was he who argued that the tyranny imposed by 51 percent of the people was every bit as threatening to democracy as the royal tyranny the colonists had fought to leave behind. In fact, many of the ideas that Guinier was advocating had been around, and respected, for literally hundreds of years.

Most persuasive of all was Guinier herself. Even in her legal writings, she had a way of making the issue of fairness personal. From her five-year-old's perspective, she argued, it was easy to see how majority rule isn't fair in many situations. If two kids want to play hide-and-seek, and three want to play tag, is it fair to always play tag?

Considering how central voting is to the American way of life, it's remarkable how little attention is paid to the ways the system really works. Like Guinier's critics, most Americans take for granted that the "winner takes all" system used in most U.S. elections is fair, if not sacred. Most people are taught that so long as the process of voting

*Temple University mathematician John Allen Paulos was one of the first to point this out.

is open and unbiased, the outcome will represent the wishes of most people, most of the time.

It is easy to show, however, that election results depend directly on the choice of voting system. Even when the preferences of the voters don't change, they can choose different winners if they change the details of the way they vote.

Since one example is worth at least a thousand words, consider the case of a typical Los Angeles family trying to decide which movie to see on Saturday night. Bombardo and Zanzibar, the eight-year-old twins, want to see *Pocahontas.* Liz-Bet, age thirteen, has been dying to see *Clueless,* while Thug, age eighteen, is dead set on *Waterworld.* Dad, Mom, and Stepmom, all forty-something, want to wax nostalgic at *Apollo 13,* but Grandma Agnes wants to try out her new pacemaker on *Species.* Grandpa Bidley and his current sweetheart vote for *Pocahontas* along with the twins.

On the face of it, *Pocahontas* wins with four votes. Using the plurality system common to most U.S. elections, the candidate with the most votes gets to make the decision for everyone else—even if, as in this case, *Pocahontas* is a minority candidate.

Dad, Mom, and Stepmom protest that the plurality system isn't fair—because more people would rather see *Apollo 13* than *Pocahontas* if those were the only options (certainly Liz-Bet and Thug wouldn't be caught dead at a Disney cartoon). So the family revotes, pitting *Pocahontas* against *Apollo 13; Apollo 13* wins.

But wait, says Liz-Bet, who's beginning to see how this voting thing works and who suspects that Granny Agnes and Thug would rather see *Clueless* than even *Apollo 13.* This time around, the family rigs its election like a tennis tournament. Pitting *Pocahontas* against *Waterworld* eliminates *Waterworld. Pocahontas* against *Apollo 13* eliminates *Pocahontas.* But *Apollo 13* loses to *Clueless.*

Now Agnes has an idea. What if she arranges the tournament so that at the last vote, only *Pocahontas* and *Species* are left? She's pretty sure that under those circumstances, *Species* would win. So she asks for a revote, and this time, in the first round, she strategically votes for *Pocahontas,* even though it's her least favorite choice. Now *Pocahontas* has enough votes to eliminate *Apollo 13*. Grandma votes for *Pocahontas* again against *Clueless*—effectively eliminating all her serious competition. When the dust settles, *Species* wins.

That example is hypothetical, but real voting paradoxes are all too easy to find. "Plurality is the worst because it has the highest likelihood of paradoxes," says Northwestern University mathematician Donald Saari. "Of all methods, it's the one where you can have the most people preferring A over B, yet B is the winner."

Plurality also means that a candidate needn't be all that popular to win. In a wide field of primary candidates, it's possible to get a winner with less than 20 percent of the vote. And if that scenario develops, says New York University political scientist Steven Brams, the winner is likely to be an extremist—"the one with the most vociferous support," not the one with the broadest appeal to the general electorate. As a result, the system can encourage extremism, reward name-calling, alienate voters, and fail to reflect the wishes of most of the people much of the time.

Mathematicians have been studying the flaws of voting systems for two hundred years. They don't agree on which system is best, but they do agree on which is the worst: It's our own hallowed tradition that says those with the most votes get to decide for everyone.

The subject is well known in academic circles. As far back as 1951, Stanford economist Kenneth Arrow proved mathematically that no democratic voting system can ever be completely fair (and later won the Nobel Prize for his efforts). This notion is known as Arrow's impossibility theorem because it proves that perfect democracy is impossible.*

Brams and others are drawn to voting systems that allow a

*Arrow didn't use undefinable terms like "fair" or "perfect democracy." However, he proved that no possible voting method could satisfy all five of the most desirable properties of a voting system.

broader spectrum of choices than simply "yes" and "no." For example, "approval voting" allows each voter to cast one vote per candidate—changing the "one person, one vote" dictum to "one candidate, one vote." That way, a voter can "approve of" as many candidates as he or she likes.

While plurality systems encourage candidates to take extreme positions that develop a hard core of support, approval voting requires candidates to appeal for broad support. As a plus, the system might do a lot to eliminate negative campaigning. "You would want to get at least partial approval from supporters of your opposition," Brams explains.

Brams argues that approval voting would also help minority candidates such as Jesse Jackson (or Pat Buchanan). "Even if Jackson couldn't win under approval voting, more majority candidates would ape his views. Minorities can't be ignored because majority candidates need their support to win."

Approval voting is not without its detractors, however. Paulos argues that it tends to produce bland, mediocre winners. "Someone who doesn't have any sharp edges won't turn people off," he says. "But sometimes you want someone who polarizes people, because one of the poles is right." Churchill, he points out, probably couldn't have won under this system.

The system Paulos likes much better is cumulative voting, in which voters can pile up several votes for a single candidate (or issue) they feel strongly about. As in approval voting, each voter has as many votes as candidates. But under cumulative voting, a voter could focus on the most important issue, instead of simply approving or disapproving of each one. (Granny Agnes, for example, could spend all five of her movie votes trying to ensure the selection of *Species*.)

Cumulative voting is already widely used in this country to elect corporate boards of directors. It's the system championed by Lani Guinier, who writes: "It is not a particularly radical idea; thirty states either require or permit corporations to use this election. . . . It is neither liberal nor conservative. Both the Reagan and Bush administrations approved cumulative voting schemes. . . ."

And if that's not enough to remove the stigma of "un-American," cumulative voting is also used in city council elections in, of all places, Peoria, Illinois.

Neither approval nor cumulative voting is "fair" enough to satisfy Saari, however. Approval voting tends to be indecisive, he argues, because it doesn't discriminate between first- and second-place choices, and cumulative voting is too complicated, in that it requires people to develop group strategies to elect winners. The system he favors allows each voter to rank each candidate as first choice, second choice, third choice, and so on. "You still need to have broad support," he says, "but it's better than approval because you designate first and second choices."

Brams and Paulos have also written reams on voting paradoxes, and true to Arrow's impossibility theorem, they still find it impossible to agree on which system is best under all circumstances. And that only scratches the surface. Other variations abound.

Until recently, however, the flaws of traditional voting systems received little attention. But today the issues that once were the province of mathematicians alone have begun to emerge into the broader political debate.

The Supreme Court has recently provided one strong incentive. The high court ruled in June 1995 that states may not take race into consideration when they draw boundaries for congressional districts. In other words, they can't carve out a district that has enough black (or Latino) voters to ensure that minorities will be at least minimally represented. (Never mind that it's always been standard practice—and perfectly legal—to carve out districts that guarantee the election of the incumbent.)

The court's decision prompted people to think more seriously about other ways of ensuring adequate representation of minority groups. The Center for Voting and Democracy, for example, helped draft legislation introduced by Representative Cynthia McKinney, the black congresswoman whose district in Georgia prompted the Supreme Court's decision. McKinney's bill would have permitted states to experiment with alternative voting schemes.

The center's director, Robert Richie, likes a system called "preference voting," a method already used to elect both houses of the Australian congress, the Irish president, and the Cambridge, Massachusetts, city council. Under preference voting, each voter ranks each candidate first, second, third, and so forth. But if someone's first-place choice seems doomed to defeat (*Waterworld*, to stick to the movie example), then that voter's second-place vote is counted instead.

While preference voting may sound complicated, Richie points out that it's used throughout the world without problems and generates far better voter turnouts than U.S. elections. It's even used by actors to choose the five nominees in each Academy Award category. "All these actors—who are not exactly rocket scientists—rank people in their field in order of preference, because Price Waterhouse said years ago it was the best way to do it." And as Richie points out, "It's not any harder than learning the rules of baseball or basketball." Despite arcane and often archaic rules, "[basketball is] a pretty good game although it's not that good for short people. The difference is, politics is the kind of game we want everyone to have a chance to play."

Each of these voting systems skews the results of an election in its own particular way. The founders of our country were aware of the need to balance the flaws in any one system through checks and balances. Indeed, the Constitution provides for a complex of voting schemes specifically designed to protect minority interests: small states are protected by the Senate's two-member-per-state rules, while populous states wield more power under proportional representation in the house; the president can rein in an out-of-control majority with a veto; and amendments to the Constitution require supermajority victories.

In the end, deciding which voting system is the fairest of them all will probably come down to, well, voting. "Before you decide on the substance, you have to decide what voting method you're going to use," says Paulos. "And then, what method are we going to use to choose the method? It can get all tangled and awful."

The one thing that most parties do agree on is that the current system is the worst. "The current system is just a disaster for most voters," says Richie, "and it's getting worse and worse."

Still, reform is not likely to come easy. Many people feel that there's something vaguely un-American about choosing voting methods based on predicted outcomes—even though the experts say that is inevitable. Common sense tells us that it shouldn't be that difficult to come up with a voting system that respects the wishes of most of the people most of the time. Common sense also tells us that a voting system shouldn't be designed with an eye to the outcome of an election—that is, designed to ensure that one side or the other should win.

Alas, these commonsense conclusions are precisely those that mathematics tells us we can't avoid. All voting systems play favorites. And even simple mathematical solutions can open up a Pandora's box of problems.

Politicians—although they seem not to remember these lessons today—learned them well during the early days of the republic. The exploding 1880 census figures, for example, prompted a revision of the ways representatives were elected to Congress. Specifically, the House of Representatives was made bigger—a change you would expect to produce greater representation for at least some states. And so it did.

However, under one scheme, the State of Alabama would have lost one seat, even though the total House membership increased. This so-called Alabama paradox forced the leaders to find a less objectionable solution. Go figure.

Chapter 10

FAIR DIVISION: THE WISDOM OF SOLOMON

> Philosophers have argued about fairness for thousands of years. What's different now is we have a formal mathematical structure. That takes it out of the ideological debate. There's science here.
>
> —*John Ledyard, economist and head of the social sciences division of the California Institute of Technology*

How should a fair society divide things that everyone wants? Who gets the house in a divorce case? How should Grandma's estate be distributed? Which countries should have the right to mine the rich minerals in the ocean floor?

The federal budget, toys in the playground, the lands under dispute in Bosnia—all have to be parceled out in a way that makes all parties content, or at least content enough not to start fighting all over again. You don't need to be a historian to see how often a supposedly "fair" settlement turns out to be a temporary solution—where the losing side simply lies low until it gains strength enough to fight again. Then the haggling—or the slaughter—begins all over again.

In 1994, however, hope appeared on the horizon. A political scientist and a mathematician seemed to have succeeded where centuries of judges, teachers, and religious leaders have failed: New York University's Steven Brams and Union College mathematician Alan Taylor came up with a new system for dividing just about anything into "envy-free pieces." Not only do all parties get what they think is a fair deal, each side thinks it got the better of the other guy.

And Brams and Taylor's work is only the tip of the proverbial iceberg. Beneath the surface, mathematics has quietly been invading political science departments nationwide with rational approaches to often highly emotional questions such as fairness. Even that bastion of hard science, the California Institute of Technology, received a $800,000 grant from the National Science Foundation to apply mathematics to what is known as "social choice theory" (in other words, fairness). Among other things, the Caltech scientists found that most people will choose to be "nice" rather than make a profit at the expense of someone else—especially if that someone else is someone they know.

For the researchers who study it, the subject of fairness turns out to be a minefield of moral and logical paradoxes. At the outset, for example, one needs to define what "fairness" really is. And that almost inevitably means different things to different people. "When my kids say something's unfair," says Ledyard, "what they really mean is, they didn't get what they want."

One of the earliest notions of fairness goes back to the Bible's King Solomon. When two women approached the king, both claiming to be the mother of the same baby, Solomon offered to slice the child in two and give half to each. As he anticipated, the baby's real mother immediately gave up her own claim on the child to save its life.

Solomon's notion of fairness survives into present-day theory. In essence, it says that fair division doesn't simply cut things into equal pieces, but considers the value various bidders place on the thing to be divided. Finding out how much people really value things is often difficult, however. Most people try to take whatever they can get rather than honestly state their choices. Solomon got around this problem by trickery: The woman who wasn't the real mother revealed the low value she placed on the child the minute she agreed to see it sliced in half. The real mother revealed her preference by offering to give up her claim.

Trying to encode the wisdom of Solomon into equations hasn't been easy, but scientists are making progress. Until recently, the problem of fair division has been tackled mainly as an academic exercise and is only in the early stages of being applied to real-life situations. Mathematicians do most of their research on a fairly simple model, called the cake-cutting problem. While the scenario may seem simplistic, its features apply to most concrete situations. It goes like this:

Suppose two people want to share a small cake. The fairest way to divide it is to let one person cut the cake in half and let the other person choose his piece first. The first person reveals his true preferences in the way he cuts the cake. For example, if he values icing more than cake, he might cut the half with more icing a bit smaller than the other half—hoping that the other person will go for the bigger piece.

Either way, both people are winners: one, because he got a bigger piece; the other because he got the most icing. Perception is as important as mathematics in making the solution work—that is, neither person gets left with a piece he didn't have a role in choosing, so neither feels envious of the other.

This same approach has been written into a law that specifies how to divide up the rich mineral wealth of the ocean floor. A first-come, first-served policy would leave developing countries empty-handed since they don't have the resources to mine the seafloor themselves. To make the division of seafloor fairer, the Convention on the Law of the Sea, an international agreement, went into effect in 1994. It specifies that the oceans' wealth be divvied up as follows: When a developed country wants to mine a portion of seafloor, it first must divide the tract into two equal shares—one for itself, and one for a consortium that preserves the land for developing countries to mine in the future. As in the cake-cutting scheme, the consortium gets to choose first.

The cake-cutting tactic works well for two players. In more complicated situations, however, three or more players often compete for equal shares.

In 1992, in response to a challenge posed in the pages of *The Sciences* magazine, Brams came up with a solution that works for three players. The first player divides the cake into three pieces; the second player is allowed to trim one of the pieces if he thinks it's bigger than the other two. The third player gets to choose first. In the end, everyone has a say, and everyone gets to choose what they think is the best piece. Ergo, the solution is envy-free.

But when Brams tried to expand his trimming system to four people, it didn't work. So he called his friend Taylor, a mathematician at Union College in Schenectady, New York. Taylor had never worked on the problem of fair division—a factor, he thinks, in helping him to come up with a radically new approach.

Taylor's breakthrough won kudos from mathematicians, because it spelled out a way to divide up anything into any possible number of "envy-free" pieces—a first. But in practice, the method is too unwieldy to use in most everyday situations. So Brams and Taylor turned their attention to a more workable method that produced similar results. Instead of viewing the goods to be divided as a cake, the new system—called Adjusted Winner—grants 100 points to each player to distribute according to their preferences.

This new system is generating excitement far beyond the mathematics community because it's simple and flexible enough to use in divorce settlements, international conflicts, and a host of other kinds of disputes.

For example, one spouse in a divorcing couple might care deeply about the family home, and so put 90 points on the house; the other spouse might care more about alimony, or tax breaks, and puts 70 points on alimony.

In the first step of this procedure, each person wins whatever they place most points on: one spouse gets the house; the other, alimony. In step two, the points are tallied to determine who has more. In this case, the spouse with the house would have 90 points to the other spouse's 70 points. So the one with 70 points might get, say, a life insurance policy rated 30 points; then the other spouse might get the family computer, worth 10.

Where there's a tie, or the points can't be divided easily, the rest is split according to a mathematical formula worked out by Taylor and Brams that's guaranteed envy-free (patent pending).

The key to eliminating envy lies in the fact that different parties value the goods differently. Each perceives that they get more than 50 percent because each puts more value on the goods they got than on the goods the other person got.

In the same way, "bads" can be divided up—for example, chores or paying taxes. Instead of awarding points to things they want, people give negative points to things they want to avoid; for some people, the price of never having to change a soiled diaper might well be worth standing in line for hours at the DMV for a car registration renewal.

Could such a system really work? It's too early to tell, since the approach is so new. However, it already has its skeptics and supporters. Some divorce lawyers are skeptical that such a rational approach could play a role in such an emotional situation. But others applaud it as a way to separate issues that can be dealt with rationally (like property and support payments) from issues that can't (like wounded pride and loss of companionship).

Caltech scientists have taken yet another approach to fairness, based on the age-old idea of auctions. Auctions, Ledyard says, counteract a person's "natural tendency to ask for more than you really need." An auction discourages people from bidding on what they don't need, because they have to pay for and take whatever they bid on—leaving them with less money for something else. In other words, they have to make choices that carry real consequences.

Of course, for an auction to be fair, everyone has to have an equal start. But not everyone agrees on what "equal start" means. For example, Caltech got involved in helping divide up responsibility for L.A.'s polluted skies. The method they came up with works like an auction, except what players bid on is license to pollute. Cutting pollution is more difficult and expensive for some companies than for others. So the Clean Air Act of 1990 allowed California companies to buy and sell licenses to pollute—permitting firms that invest in expensive technology to recoup their expenses by selling unneeded pollution allowances to others.

In order to make the process fair, however, the total amount of pollution permitted had to be fairly divided to start with. "In this case, equal is not fair," says Ledyard. "It's not fair to give the same allowance to Joe's Bar and Grill as to SoCal Edison."

Instead, the allowances granted by the air quality board are based on levels of pollution during the years 1989–1990. Based on those figures, a factory might receive a permit to spew out three thousand pounds of pollution per year; a donut shop might receive an allowance of fifty pounds. But as Caltech economist Colin Camerer points out, that system "rewards people for polluting in the past."

The system, in other words, is not perfect. But at least it's a start at inventing a method that gives incentives to everyone to keep pollution down—and save money in the process.

Further complicating attempts to divide things fairly is the human factor. The NSF grant to Caltech was designated mainly for experimental testing to determine how well real people's behavior matches the mathematical models.

Researchers have assumed, for example, that people would always act in a way that would maximize their gain. However, Caltech's Camerer has found in a series of experiments that good manners can get in the way of rational behavior—especially if players were not strangers. They'll give up tangible rewards, even cold, hard cash, in order to avoid appearing greedy or selfish. People further alter their economic behavior when they are in personal contact. For example, people voluntarily contribute to communal goods such as streetlights and holiday parties. "But if they are face-to-face, the rate of contribution goes way up," says Ledyard. "Why is an open question."

Finally, the choices people make in one-to-one situations can be quite different from the choices they might make as a member of a larger group or community: for example, in a situation where a community is trying to decide whether a tax break is fair compensation for having a chemical factory or hazardous waste dump in its backyard.

Rakesh K. Sarin of UCLA studies how people weigh relative risk (say, of living with toxic fumes) against possible gain (say, more jobs). He's found what an individual might perceive as a good deal from his personal perspective might be unacceptable to him as a part of a group. The individuals might perceive the trade-off as perfectly fair. "But the group may have an outcry," he says, even though the group is composed of these same individuals.

The main benefit of this new mathematical attack on fairness, says Ledyard, is the potential for creating new kinds of social systems more or less from scratch. Until now economists and political scientists have concentrated almost exclusively on the study of existing systems, such as elections or financial markets. "We would look at auctions and elections and ask: Are they fair? It's like studying rabbits and mice as they arise in nature."

Now, he says, they can turn the process on its head, first getting agreement on what fairness is (for example, it has the property of being "envy-free") and then designing a process to deliver it. "So instead of studying elephants," he says, "you're playing with DNA to create elephants."

One might call the system of auctioning pollution allowances one of the first of a new species of social institutions to be created from the ground up. If it catches on, who knows what the future will hold? Similar approaches could be applied to everything from deciding who gets dorm rooms to who gets called up for military service.

At the very least, the social scientists and mathematicians are learning how to step around the paradoxes inherent in most approaches to fairness. "When someone says they want a situation to be fair," says Ledyard, "we now understand that opens up a huge range of possibilities."

Chapter 11

THE MATHEMATICS OF KINDNESS: MATH PROVES THE GOLDEN RULE

Surprisingly, there is a single property which distinguishes the relatively high-scoring entries from the relatively low-scoring entries. This is the property of being NICE. . . .

—*Robert Axelrod, in* The Evolution of Cooperation

Life did not take over the globe by combat, but by networking.

—*Lynn Margulis and Dorion Sagan in* Microcosmos

"Do unto others as you would have others do unto you." "An eye for an eye." "Get it while the getting's good." "He ain't heavy, he's my brother."

Selfishness and altruism have always been uneasy partners in human affairs. Churches and scout troops exhort us to lend a helping hand to those in need; at the same time, advertisers and politicians encourage us to be as greedy as humanly possible. Indeed, the idea that greed is all to the good has become encoded in a kind of religion of U.S.-style capitalism: The more you're out for yourself, the better off the whole society will be.

This win-at-all-costs strategy gains strength because it appears to be founded on Mother Nature's own laws. Charles Darwin's idea of survival of the fittest suggests that only the meanest, most competitive, most selfish individuals will make it to the top of the evolutionary heap. Compromise, cooperation, and kindness are for losers and wimps. In a capitalistic society, failure to be selfish is akin to economic treason.

For a long time, people have accepted this philosophy as undeniably true. But for the past two decades, mathematicians have been studying survival strategies to find out which are truly best. To almost everyone's surprise, they have found that nice guys can and frequently do finish first. In tournaments designed to pick out winners in a variety of conflict situations, the top dog turns out to be not the most ferocious but the most cooperative. Ironically, the strategies that have emerged from the mathematical research sound a lot like old-fashioned homilies: think ahead, cooperate, don't covet your neighbor's success, and be prepared to forgive those who trespass against you.

Much work in game theory has focused on one of the most unsettling paradoxes of all, the so-called prisoner's dilemma. It's usually explained as a familiar cop show scenario. Two partners in crime are kept in isolated cells. Each is told that if he blows the whistle on the other, he might be able to go free. If he remains mute, each prisoner knows, the authorities might not have enough evidence to convict him—unless, of course, the other prisoner rats on him first. Which strategy works best—keep silent or strike a deal?

Variations on this theme, I think, make the inherent paradox even clearer. Assume, for example, that you've outgrown your old car but desperately want a family sailboat for Sunday afternoon excursions. Another person—whom you contact through the newspaper—desperately needs a car like yours and has exactly the sailboat you crave—a boat she no longer uses. You both agree that a swap would be a fair trade.

Now assume for some reason that the trade needs to be kept secret. You both agree to put the car/boat in predesignated places. The problem is: What happens if you leave behind your car, and the boat isn't where it's supposed to be? You've been cheated!

The boat owner faces exactly the same dilemma.

Logically, you might add up the pros and cons this way: If you leave the car, but the other person doesn't leave the boat, you get robbed. If you don't deliver the car, and she doesn't deliver the boat, then you come out even. If you don't leave the car and she does deliver the boat, you get something for nothing.

Logic points you to an inescapable conclusion: No matter what the other person does, you're better off not leaving your car. The other person's logic leads her to the same destination. Outcome? Neither of you gets what you wanted.

A prisoner's dilemma pops up anytime going after your own immediate interests results in disaster if everyone does it. Should you throw your trash out the car window, or wait until you find a garbage can? Listen to public radio for free, or pay your way? Abide by disarmament agreements, or cheat and hide your arsenals?

Clearly, if one party cooperates while the other cheats, the cooperator is a sucker. But if both cheat, no one gets anything.

Looking in on the situation from the outside, it may be clear that cooperation is a better tactic for both sides. But from an individual player's point of view, there's always temptation to try to get the better of the other guy; you always have a chance of winning more by *not* cooperating.

Why, then, do people cooperate at all? This is the question that intrigued political scientist Robert Axelrod of the University of Michigan in the 1980s. If dog eat dog is the law of the jungle, why is cooperation so common among humans and other species as well? During trench warfare in World War I, Axelrod points out, soldiers on opposite sides of the front lines formed tacit agreements to live and let live—in direct defiance of orders from commanders. An officer in the British army, writing in his diary in August 1915, recounts how after a Prussian shell exploded in the British camp (during teatime, no less), a German soldier climbed out of his trench and crossed no-man's-land to apologize: "We hope no one was hurt. It is not our fault, it is that damned Prussian artillery."

Closer to home, it's not even clear why people obey traffic signals. Individually, there's not much motivation for stopping at red lights—short of the very off chance of being caught. Yet, most of the time, people do it anyway. They leave tips for waiters they may never see again, pick up after themselves when no one is looking, show kindness to total strangers.

To try to resolve the paradox, in 1980 Axelrod invited experts in game theory to a tournament of repeated games of prisoner's dilemma. Each entrant would submit a strategy, and the various strategies played against each other by means of computers. Points were assigned to outcomes and tabulated.

To almost everyone's surprise, the most successful strategy turned out to be an ingeniously simple program created by Anatol Rapoport at the University of Toronto. Called Tit for Tat, the program's first move is always to cooperate. After that, it simply echoes whatever its opposition does. If the opposition cooperates, Tit for Tat cooperates. If the opposition defects, Tit for Tat retaliates in kind.

In this sense, Tit for Tat embodied both biblical injunctions: an eye for an eye, and the Golden Rule. Or as William Poundstone sums it up in a book about classic game theory problems, the program's message is: "Do unto others as you would have them do unto you—or else!"

By not ever being the first to defect, Tit for Tat was what Axelrod called a "nice" program. As it turns out, most of the winners in computer simulations that Axelrod has run have been nice; most of the losers were not. Tit for Tat could also be forgiving—that is, even after the opposition defected, Tit for Tat would occasionally give cooperation another try. The lesson, says Axelrod, is "be nice and forgiving."

It is also important to be clear. Very complex computer programs fare no better than random ones in such simulated games because no one can figure out what their strategy is and respond in kind.

Axelrod then held a follow-up tournament. This time he got entries not only from game theorists, but also from researchers in biology, physics, and sociology. And this time, everyone knew about the success of Tit for Tat and other "nice" strategies. Nevertheless, Rapoport's simple program won again. The other experts, Axelrod concluded, all "made systematic errors of being too competitive for their own good, not being forgiving enough."

In a final round, Axelrod wanted to see what would happen if he pitted all the programs against each other in a kind of Darwinian evolution, where survival of the fittest meant success for those who produced the most viable offspring in the next generation—the number of offspring being determined by the number of points.

This time, Tit for Tat did well, but so, at first, did some very cutthroat, exploitative strategies. Then a funny thing happened: The exploitative strategies ran out of prey. There was no one left to gobble up. As Axelrod puts it, "In the long run [a strategy that is not nice] can destroy the very environment it needs for its own success."

The tournament also had lessons for the envious. If one strategy envied another's success and tried to do better, it would usually wind up cutting off its nose to spite its face. That is, the only way to get the better of an opponent would be to attack, and that would set off another round of nastiness that would make everyone worse off.

"There is no point in being envious of the success of the other player," says Axelrod, "since [in this kind of game] the other's success is virtually a prerequisite of your doing well yourself."

A final requirement for success was a stable, long-term relationship, where the same opponents would play each other again and again. In such a situation, it paid to be cooperative. This explains the relationship of the World War I soldiers, who faced each other month after month, or people in tight communities, or national leaders who need each other in an ongoing series of negotiations.

More recently, New York University's Steven Brams made great strides in making game theory more realistic.* While he was at it, he got interested in whether it might be possible to use mathematics to model human emotions—and therefore come up with strategies for getting out of frustrating situations.

In his "frustration" games, one player is stuck in a bad position, while another player is satisfied and has no incentive to change his tune. The first player can't get out of the situation without also hurting himself.

An example might be a family with an unruly teenager who refuses to follow any of the parents' rules. The parents don't want to become too Draconian, because that might hurt them, too. (Say the teenager uses the family car to take his little sister to school, and the parents would lose that service if they take away his keys. Or say they impose a no TV rule, but that means they must give up their favorite shows as well.)

If the parents get frustrated enough, however, they might be willing to hurt themselves (at least temporarily) simply to break out of the deadlock.

Another recent and socially relevant twist on game theory illuminates the effects of obvious labels on players—like skin color or na-

*Details can be found in his book, *Theory of Moves.*

tionality or gender. As described by Poundstone, this variation on Tit for Tat changes the rules slightly. Players would always cooperate with other players of the same group, but not with players bearing different labels. Thus, males, or blues, would always cooperate with other males, or blues, but not cooperate in encounters with females, or reds.

Not surprisingly, in this game of Discriminatory Tit for Tat, the majority group always did well, but minorities did very badly. The reason is not difficult to figure out: Where majorities had most of their daily encounters with others of their own kind, and thus were treated "nicely," minorities were forever bumping into their opposites, who would always "defect," or fail to cooperate.

It's possible, Poundstone concludes, that such behavioral dynamics could account for the compelling allure of minority communities—be they religious, racial, or even financial. Even a "ghetto," in this sense, can be "a safe haven where most interactions are likely to be positive."

Curiously, evidence from the world of the living—that is, biology and genetics—seems to confirm some of the "abstract" arguments to come out of game theory. If these notions are right, then the evolution of species has depended a great deal less on "dog eat dog" and a lot more on "dog learns to live cooperatively with other dogs" (not to mention humans) than anyone imagined.

Just because the "fittest" tend to survive, in other words, doesn't necessarily mean the "fittest" are the strongest, or meanest, or even the most reproductively profligate; the fittest may be those who learn best how to use cooperation for their own ends.

Controversial microbiologist Lynn Margulis has vastly extended the idea that symbiosis (living together without the benefit of clergy, one might call it) has been a major force in shaping organisms. From trees to fish to fungus, all kinds of living things take nourishment from each other, build communal housing, use each other, and generally form all sorts of lifelong partnerships and odd arrangements for the mutual benefit of all concerned.

Margulis has suggested that the cell itself arose from such cooperative arrangements among subcellular beings. Cells are packed full of specialized components that metabolize food, produce and store energy, propel the cell, shape its internal structure, and so forth. A good deal of evidence already supports Margulis's idea that cells are more like colonies of cooperating individuals than survivors of some fierce competitive race to "success."

Other biologists have argued—on a variety of different bases—that there is probably a gene for altruism and that humans (not to mention ants and bees and other intensely communal creatures) carry it within them as part of their genetic baggage. Altruism, wrote the late Lewis Thomas, "is essential for continuation of the species, and it exists as an everyday aspect of living."

After all, it's well known that creatures as various as vampire bats and stickleback fish put their own lives at risk to feed their fellows—even when the fellows happen to be unrelated.

In his usual lyric way, Thomas fashioned these facts into a lesson of near-biblical proportion:

> I maintain that we are born and grow up with a fondness for each other, and that we have genes for that. We can be talked out of that fondness, for the genetic message is like a distant music, and some of us are hard-of-hearing. Societies are noisy affairs, drowning out the sound of ourselves and our connection. Hard-of-hearing, we go to war. Stone deaf, we make thermonuclear missiles. Nonetheless, the music is there, waiting for more listeners.

He may well be right, but the living genes aren't the only ones hearing the music. Carl Zimmer wrote in *Discover* magazine about a computer whiz named Maja Mataric of Brandeis University. She managed to get fourteen robots to cooperate in such simple tasks as retrieving a puck. Remarkably, cooperation wasn't a talent that she programmed into them. They learned it themselves. Instead of all ganging up on the same prize at the same time, she programmed them to pay attention to what the others were doing. Within fifteen minutes of practice, they acquired a taste for altruism.

What all this says about robots or vampire bats or even mathematicians, I'll leave to further study. Even if cooperation didn't steer human evolution, it probably wasn't completely absent from the picture, either. Perhaps the mathematicians' study of human interaction will someday help point the way out of what seems to be humanity's increasingly common lament—or as Rodney King puts it: "Why can't we all just get along?"

PART IV

The Mathematics of Truth

Nature knows what she is doing, and does it,
even when we cannot find out.

—*Sir Arthur Stanley Eddington*

This longing for absolute truth has to do with religion,
not science.

—*Lorraine Daston, historian of science at the University of Chicago*

Why do we age? Why do some people get struck by lightning? Why do others have all the luck? Why does a bridge collapse in an earthquake? Why do so many people believe in the paranormal? Why does a tossed coin come up heads half the time, and tails the other? Why can't time go backward?

OK, so maybe it's a bit pretentious to talk about the mathematics of truth. But mathematics offers some powerful ways to get at least closer to the truth—and there are methods in use almost everywhere we look, although most people are not aware of them.

For example, one sensible way to find solutions to problems is finding out what caused the problems in the first place. And without mathematical tools, finding causes would be even more difficult than it is.

But getting to the truth about causes can be a very tricky business. Often, causes are tangled up with hidden connections and influences. Sometimes the "cause" is chance. This bothers us because we seem to have a strong urge to root out the fundamental connections between cause and effect. We hate the idea that events are happenstance. Happily, however, cause and chance are not mutually exclusive. They merely have a more complex relationship than people previously thought. The first chapter in this section discusses two connected ways of quantifying the relationship between cause and effect—probability and correlation.

The second chapter deals with how truth can be proven. During the time I was writing this book, the O.J. Simpson case was winding its way through court(s), a subatomic particle called the top quark was discovered, and an ancient mathematical problem known as Fermat's last theorem was proven true. Curiously, the lawyers, scientists, and mathematicians used much the same kinds of tools to prove their cases. And all of the tools, at least to some extent, came from the pantry of mathematics. All are legitimately recognized tests of validity in science, none of them definitive:

- *Probable truth relies on statistical arguments to weigh which of several possibilities is more likely to be true. When something must be proven beyond a reasonable doubt, just what does that mean? What level of doubt is acceptable? One in twenty? One in a trillion?*

- *Many legal and scientific arguments rely on logical truth, which is based on the belief that following clear rules of deduction always leads to unambiguous conclusions. If* x, *then* y. *Logical arguments are supposed to make sense. Truth is supposed to emerge through the power of reason.*

- *Eventually, lawyers have to present their cases before juries of peers—just as scientists submit their work to peer review. This is an example of truth by consensus, in which informed people (experts or juries, or both) evaluate evidence and come to an agreement. Frequently, these agreements are temporary coalitions, later overruled by new insights or information. Scientific truth, like legal truth, is less a collection of facts than a running argument. The difference is, legal truth has to come to a conclusion quickly.*

The final chapter looks at one of the most powerful guides to truth of all—the idea of symmetry. In effect, exploring symmetry allows one to pick out those aspects of nature that are truly fundamental. The way to get to these unvarying truths, perhaps ironically, is to become acutely aware of one's own point of view. Einstein's theories of special and general relativity are case studies in how particular frames of reference can be quantified, and thus penetrated, to discover deeper truths.

Chapter 12

THE TRUTH ABOUT WHY THINGS HAPPEN

PROBABLE CAUSES: THE ROLL OF THE DICE

A very slight cause, which escapes us,
determines a considerable effect which we cannot help seeing,
and then we say this effect is due to chance.

—Henri Poincaré

Probability exerts a peculiar fascination
even over persons who care nothing for mathematics.
It is rich in philosophical interest and of the highest
scientific importance. But it is also baffling.

—James R. Newman, in The World of Mathematics

The roll of the dice is the epitome of happenstance. But there is no reason why it should be so. Dice, like planets, follow the laws of nature. They are pulled toward the table by the predictable pull of gravity, slowed as they tumble through the air by the well-understood electrical stickiness of matter, spun in circles obeying the same rules of momentum conservation that twirl planets and toppled Tonya Harding.

There is nothing left to chance in this drama. Yet chancy it is, time after time. You can toss a penny a million times and, on toss number one million and one, the coin still has a fifty-fifty chance of turning up heads or tails.

Newman's musings on probability are now fifty years old, but probability is as baffling as ever. How remarkable it is, he writes, that despite the attention bestowed on this science and its enormous influence, mathematicians and philosophers are quite unable to agree on the meaning of probability. Their disagreement is less easily explained than that of the three blind men describing an elephant. For in this case the observers are not blind and the creature is of their own design.

Chance is both tempting and troubling. It tempts in that it excuses us from all responsibility: what happens by chance is exempt from rhyme or reason; it is something we can't know and can't control; it falls out of the blue or comes out of nowhere.

And it troubles for all the same reasons.

Things that happen by chance are effects in search of causes. They do not belong in a reasonable world. To some, including Albert Einstein, probable causes were simply unacceptable. Even people who know nothing about the great late physicist have no doubt heard some version of his famous lament to fellow physicist Max Born: "You believe in the dice-playing god, and I in the perfect rule of law. . . ."

Einstein was upset by the new view of the atom, which revealed a subatomic realm of rampant uncertainty. Ever more sensitive experiments with light and matter in the early part of the century drove physicists to a seemingly inescapable conclusion: At subatomic scales, particles could not be pinned down to a particular place at a particular moment of time. Their behavior could not be predicted at all, except in terms of probabilities. This new understanding of the atom had the power to explain a score of outstanding mysteries. Yet Einstein was unpersuaded. Even these obvious successes, he said, "cannot convert me to believe in that fundamental game of dice."

Whatever to make of subatomic particles that behave as unpredictably as dice? Are they not propelled by causes? Do they traipse around willy-nilly without pattern? Motivated by no force? Stumbling here and there like a group of drunken partygoers?

Unpredictability, argued Born, does not mean that subatomic behavior does not have causes; it only means that it has causes too subtle and complex for us to untangle. (The full drama of the Born-Einstein correspondence is contained in Born's *The Natural Philosophy of Cause and Chance*—highly recommended to anyone inclined to pursue this still delicious debate.)

Today most physicists have accepted the curious fact that there's no real contradiction between probability and causality. Indeed, all physical forces (including familiar ones like electricity and gravity) are described in terms of probabilities. The so-called strong force that holds nuclear particles together inside an atom is a kind of interaction among particles that is orders of magnitude more likely to happen than the so-called weak force involved in radioactive decay.

"The atomic world of small action," writes physicist Philip Morrison, "is ruled by a fusion of cause and chance." This comes as a great relief to those who feel a completely predictable universe suffocates creativity and free will, and yet who feel that a completely unpredictable universe is too crazed to even contemplate.

A universe that can be predicted only in a statistical sense allows room for order *and* chance. "There is room to breathe in such a world," writes Morrison. "Yet it is no world of caprice or chaos. Chance and cause have been wonderfully married into a point of view in which precise pattern governs potential events, and yet in which the variety of potentialities allows the full growth of that novelty which we know to govern the world we live in."

In some sense, it's a matter of scale. As Ivar Ekeland points out in *The Broken Dice*, the difference between what's random and what's deterministic is partly dependent on dimension. Billiard balls, for example, are routinely invoked as the model of predictable physics. No one calls billiards a game of chance. All it requires to get the ball in the pocket is a well-measured poke of the perfectly positioned cue.

Dice, on the other hand, is the epitome of chance. Yet billiard balls and dice are both governed by the same natural laws. "So in the final analysis, chance lies in the clumsiness, the inexperience, or the naïveté of the thrower—or in the eye of the observer. In fact, one might perfectly imagine a civilization in which the rolling of the dice would be a sport and billiards a game of chance," writes Ekeland.

The dice would be much bigger, of course—about the size of billiard balls. One could throw them much like boccie balls—or bowling balls; billiards would be smaller, like dice, and subject to more obstacles—say, like a game of pinball, which is a kind of billiards clearly based on chance.

Either way, the distinction is neither clean nor clear.

Indeed, there is a very real sense in which chance is a cause. Take the roll of the dice, for example. Any Las Vegas gambler had better know that two dice rolled randomly come up seven more often than any other number. The reason is not mysterious: Given that each face of each die has an even chance of landing on top, there are more ways to make seven than any other number. To make a 7, the roll of the dice can produce the pairs 6 and 1, 5 and 2, 4 and 3, 3 and 4, 2 and 5, 1 and 6. There is only one way that two dice can add up to 12, however, or 2.

Seven, in other words, has more opportunities to turn up than any other number. Thus seven is the most likely, most probable, result. Probability is the land of opportunity. (Ask any life insurance salesclerk.)

Of course, the gambler does not know which die will turn up 5 or which will turn up 2 any more than the life insurance clerk knows which person will die next. But put a lot of dice or people together, and the pattern becomes perfectly predictable. Chance becomes certitude.

But due to the duality inherent in its nature, probability is also deeply mired in ignorance. A hundred years ago, Henri Poincaré put it this way: "Fortuitous phenomena are, by definition, those whose laws we do not know."

Newman concludes that the whole concept of chance is only a euphemism for ignorance. And yet, he allows, chance is also marked by regularity, "an order within disorder." Today, we know that probability can be found lurking at the bottom of many kinds of truth.*

It's the uncanny regularity of chance that makes the behavior of subatomic particles, large crowds, and traffic accidents perfectly predictable. Sooner or later, the things we call coincidences always happen, in a surprisingly predictable way.

In New York recently, someone won the million-dollar lottery for a second time. A lucky charm at work? An auspicious arrangement of stars? "That's really not all that surprising," says Chuck Newman, a mathematician at the Courant Institute in New York. "After a period of ten or fifteen years, when you have two winners a week, you have thousands of previous winners running around and buying tickets."

This is why statisticians distrust studies that show, for example, unusual rates of cancer from people who live under power lines. Out

*See Probable Truth in Chapter 13, "The Burden of Proof."

of a thousand communities, it is perfectly natural to find a clustering of cancer occurrence in some of them, just as it is perfectly natural for the coin to come up heads ten times in a row. It might suggest a connection between power lines and cancer, but it certainly doesn't prove one. The mathematics of probability is an essential tool for sorting out what might be expected to happen purely "by chance"—which in no way implies that it doesn't happen for a reason.

As Newman suggests, probability is baffling. It's Janus-faced, both predictable and not. This puzzling duality is behind the frustrating inability of relatives of lifelong smokers to sue cigarette companies for their wrongful deaths. That cigarette smoke causes cancer has been proven beyond a reasonable doubt. But that smoking caused any particular case of cancer is near impossible to demonstrate.

Probability makes a pretty good crystal ball for looking into the future as long as the numbers are large enough. Once again, scale changes everything,* and the "laws of chance" become as reliable as any other laws of nature.

Humpty-Dumpty was fated to break into a dozen pieces when he fell from the wall because of chance. Bodies and engines wear out because of chance. Files get messy and ice cubes melt because of chance.

On an atom-by-atom analysis, there is no natural law that says that every slow-moving (therefore freezing-cold) water molecule couldn't suddenly crowd into, say, the northeast corner of a glass to create a cube of ice. It could happen. It doesn't because there are many millions of pathways for the slow molecules to follow as they spread throughout the liquid—and very few pathways that lead them all into a small cube of space in the northeast corner.

Disorder is far more probable than order, so disorder happens; disorder is an opportunistic disease, and it is highly predictable in

*See Chapter 5, "A Matter of Scale."

the long run. Sooner or later, the ordered crystal lattice of the ice cube, the eggshell, the elasticity of skin, all tend toward a state of inevitable disintegration. It is probability that makes time a one-way street.

Still, we conclude that probable causes aren't somehow as real as physical causes. When the National Rifle Association says, "Guns don't kill people; people kill people," they are falling into this logical trap. The mere fact that guns are ubiquitous makes their use more probable. To argue otherwise is to argue that a thousand-piece jigsaw puzzle has no more chance of falling into a disorganized heap when it topples from the table than does a two-piece jigsaw puzzle. It isn't impossible that all the pieces could arrange themselves in the proper order purely by chance. It isn't impossible that a heavily armed society can avoid a high murder rate. It's only highly, highly improbable.

Or as the authors of *The Empire of Chance* sum it up: "To be a cause generally means to make a difference: in the hypothetical situation where the cause is absent, the effect would not have occurred."

This does not mean, however, that the very improbable never happens. Given eons enough, you could flip a coin enough times to get a million heads in a row. Eggs would unscramble; perfume would steal silently back into the bottle. Of course, the time required would be far longer than the age of the universe, but that's just a practical impediment.

Newman asks: How long will it be before a book jumps UP into your hand? The answer, he reminds us, is not never. "The right answer is that it will almost certainly happen sometimes in less than a googolplex of years—perhaps tomorrow." (A googolplex is a number called a googol multiplied by itself a hundred times, and a googol far exceeds the number of particles in the universe.) You only need to wait, he says, "for the favorable moment when there happens to be an enormous number of molecules bombarding the book from below and very few from above. Then gravity will be overcome and the book will rise."

Miracles occur. Probability can be overcome, but it is not easy. Either you must wait until eternity. Or you must do work. It takes energy to breathe order into chaos. You must paint the house, alphabetize the files, pump electricity into the refrigerator to make the ice cube, exercise to keep the muscles toned.

Life performs this magic all the time, taking scattered sunlight and dirt and water and turning it into corn and tortillas and burgers. Ashes eventually go to ashes, and dust to dust, but in between are people, puppies, and begonias. Life is highly improbable, and yet it happens. It requires reversing the inevitable disintegration of forms, the very arrow of time.

Life goes against the grain. Each new child created diminishes the disorganization of the universe—a little at a time. Michael Guillen describes how Rudolf Clausius came upon the equation behind the inevitable disorder of the universe, the unlikeliness of life, in the mid-nineteenth century. "Like all unnatural behavior [he discovered], life was the result of some engine whose coercive efforts were able to reverse the laws of normal behavior—the way a refrigerator was able to make heat flow from cold to hot." (Heat normally flows from hot to cold.)

Clausius's newfound understanding, says Guillen, "was the first scientific explanation of why everything in the universe aged and eventually died." Having discovered that the amount of disorder in the universe is always increasing (due to disorder's overwhelming probability), Clausius understood why "Death always outscored Life, which explained why each and every life always came to an end. Always."

"So the improved Newtonian Universe must cease and grow cold," says Thomasina's tutor Septimus in Stoppard's *Arcadia,* suddenly realizing the import of his student's mathematical discovery that disorder is the inevitable direction of things. "Dear me."

"Yes," Thomasina replies, "we must hurry if we are going to dance."

CORRELATION AND CAUSE: SMART KIDS WITH BIG FEET

> Correlation doesn't tell you anything about causation, but it's a mistake that even researchers make.
>
> —*psychologist and statistician Rand Wilcox of the University of Southern California*

One popular method for getting to the truth about causes is correlation—that is, figuring out which things happen together. In a sense, this is the mathematical version of the old saying "Where there's smoke, there's fire." If smoke and fire usually happen together, then we can assume that one causes the other. Maybe.

To be sure, making statistical correlations has led directly to a great many important truths. For example, the link between cigarette smoking and lung cancer was seen as a correlation long before any causal mechanism between smoke and cancer was understood. The same is true of many environmental hazards, including DDT. The rise in carbon dioxide levels in the atmosphere since the start of the industrial revolution, some researchers think, is correlated so strongly with an increase in global temperatures that a climate catastrophe must surely be at hand.

Mathematicians, however, know that correlation does not necessarily mean causation. As John Allen Paulos points out, big feet are strongly correlated with high math scores among schoolchildren. The reason is not that big feet mean big brains, but rather that older children tend to have acquired both bigger feet and more math instruction than younger children.

Correlation means only that one thing has a relationship to another. Ships rise with the tide because the water level pushes up the ships. But do skirt lengths rise and fall with stock prices? And if so, can a causal relationship be found?

Often, correlations are nothing more telling than coincidence or timing. Studies routinely reveal a strong statistical link between divorced parents and troubled adolescents. But it is also true that adolescents are attracted to trouble no matter what parents do. One could just as easily argue that orthodontia causes puberty, since the two seem to occur at roughly the same time.

The same is true of the effectiveness of almost any imaginable cure for the common cold. A solid statistical case can be made that the majority of people who take cold medications feel significantly better after a week's time. On the other hand, since colds tend to cure themselves within the space of a week, almost any other remedy could prove equally beneficial—from Mom's chicken soup to watching old episodes of *Star Trek.*

"Some of the correlations are just plain silly," says Wilcox. "You can prove that the amount of time you spend practicing tennis correlates with sunburn. But that doesn't mean that tennis causes sunburn." The tennis-sunburn link is a rather obvious example of a correlation masking a hidden effect—in this case, that tennis is normally played outdoors, where the Sun is most likely to be found.

But equally silly correlations find their way into all kinds of seemingly well-reasoned arguments. For example, Hal Lewis, in his book *Technological Risk,* cites the case of the antifluoridation publication that warned that most cases of AIDS were found in cities with fluoridated water. "It could just as easily have said that they occur in cities with a public library," writes Lewis. "Equally true—equally irrelevant."

Studies that have tried to link IQ with race are routinely exposed for relying on much the same fallacies. It's possible to make a strong correlation between race and IQ, for example, without knowing anything about the true causes involved. For one thing, almost anything that correlates with high IQ is also associated with high income. This conclusion comes as no surprise, given that affluent parents can more easily afford better schools, more books and computers, and generally raise more healthy, better-nourished children. Studies of IQ and race, experts say, may mask the stronger relationship between white skin and wealth. "It's quite possible that two things move together, but both are being moved by a third factor," Stanford statistician Ingram Olkin says.

Income and good health tend to rise and fall together in much the same way. The strongest link between electrical power lines and cancer may well turn out to be the high probability that poor people live and work in areas close to power lines.

Medical studies are minefields of correlations that may or may not be meaningful. Mark Lipsey of Vanderbilt University studied the relationship between alcohol use and violent behavior. "People believe that alcohol is causative," he says. "But the research base is not adequate to support that conclusion. It may be that the same people who are prone to violence are prone to alcohol abuse."

Causes and effects can even get reversed. Take the correlation between fitness and exercise. Some studies showing that exercise makes people fit were later challenged by other studies suggesting that fit people simply like to exercise more because they feel better—not vice versa.

The definitive studies in the environment versus genetics argument (nature versus nurture) are often taken to be a series of twin studies in which researchers claimed to find that identical twins reared by different sets of parents were remarkably similar; they smoked the same brands of cigarettes, leaned toward the same political views.

Yet genetics may not be the main reason that identical twins raised apart seem to share so many tastes and habits, says Richard Rose, a professor of medical genetics at Indiana University. "You're comparing individuals who grew up in the same epoch, whether they're related or not," says Rose, a collaborator on a study of sixteen thousand pairs of twins. "If you asked strangers born on the same day about their political views, food preferences, athletic heroes, clothing choices, you'd find lots of similarities. It has nothing to do with genetics."

Comparing more than one factor always complicates the issue. When dealing with income, age, race, IQ, and gender, the effects of these covariants, as statisticians call them, can be almost insurmountable. Impressive-sounding statistical methods such as multiple-regression analysis are said to eliminate this confusion by controlling for certain variables, erasing their effects. To see what effect shoe size really has on math scores, for example, you might control for the influence of grade level; only a comparison of children in the same grade would be meaningful. But mathematically erasing influences that shape life as pervasively as race, income, and gender is far more difficult, researchers agree. "There are lots of ways to get rid of [these variables]," Wilcox says. "But there are also a million ways that [the methods] can go wrong."

Comparing groups—blacks and whites, or boys and girls—muddles the matter even further. Differences within groups are frequently far greater than differences between groups. That is, when people say that certain jobs should be reserved for men because they are, on average, stronger than women, that ignores the millions of women who are stronger than millions of men. Imagine Roseanne getting turned down for a job as, say, a forklift operator, because women aren't as strong as . . . Pee-wee Herman?

The same is true of IQ scores of blacks and whites. The variation among blacks is far greater than the difference between blacks and whites. In fact, the curves plotting IQ against race mostly overlap.

Total Payroll	$1,977,500	Other Workers	7,500
"Average" (Mean) Salary	131,833		5,000
			5,000
Boss	1,000,000		5,000
Vice Boss	500,000		5,000
Managers	200,000		5,000
	200,000		5,000
Senior Workers	10,000		
	10,000	Median Salary	10,000
	10,000	Mode	5,000
	10,000		

That means that the percentage of blacks who outscore whites is substantial.*

Comparing groups is also tricky because you can't compare everyone in the one group with everyone in another. Therefore, most studies compare averages. And "average" is about the slipperiest mathematical concept ever to slide into popular consciousness.

Let's say the payroll of an office of fifteen workers is $1,977,500, and the boss brags that the average salary is about $131,833 (see chart). That's assuming everyone shares equally. But what if the boss takes home $1 million, pays her husband $500,000 as vice president, and pays two other vice presidents $200,000 each? That means the average salary of the other workers is far less—about $7,045. Yet nothing is technically wrong with the math. Rather, something is wrong with the choices of "average."

In this case, using the average known as arithmetical "mean" (dividing the total by the number of workers) disguises gross disparities. The median (the salary of the person in the middle of the range of employees) would provide the more realistic average—$10,000. One could also use the mode, or most common number on the list—$5,000.

*See Chapter 4, "The Measure of Man, Woman, and Thing."

Take a statistic almost everyone has heard: that the average length of a marriage is seven years. But what if the kind of "average" you're using is a "median"? The median value is simply that number that falls in the middle. In a sample of 1,000,000 marriages, 499,999 could end within one year, and another 499,999 could end after fifteen years. If two marriages ended after seven years, seven would still be the "average," if the average is a median.

The special property of the numerical beast known as a bell curve is that mean, median, and mode all fall in the same place, so it shouldn't make much difference which you choose to compare two groups. Discovered several centuries ago, the curve graphically shows a range of numbers that corresponds surprisingly well with any normal variation: IQ scores, measurements of height, how many times a coin comes up heads or tails, the age at which people die, whatever. Just as most people die around 70, and very few at 40 or 100, so the size of, say, birthday cakes will cluster around a twelve-inch diameter, with cupcakes and wedding cakes taking up the extremes.

The extreme results fall at the tails of the curve, and the most common fall at the top. The same is true of test scores: most people score in the middle, with a few (Forrest Gump and Albert Einstein) dangling out on the ends. But bell curves are rarely perfect. A single unusual person can have an unusually large impact. Even a few IQs that hang way out at the low end of the heap—or a few high ones—could skew the entire picture.

Another pitfall of comparing groups is defining who qualifies for membership. Who is "black," for example, and who is "white"? Most people, we now know, are mixtures. An amusing consequence of fuzzy thinking in this matter popped up in *The Bell Curve.* The book claims that Hispanics have lower IQs on average than Europeans, and Europeans in turn have lower IQs than Asians. But University of California, Davis, economist Thomas W. Hazlett points out that

"Hispanics" evolved from a merging of two groups: Europeans and American Indians. And American Indians themselves emigrated "from . . . Asia," according to Hazlett. "Oops!"

Hazlett suggests it might be more instructive to compare IQs of "people who worry about racial IQs versus those who do not care."

Recently, statisticians have discovered yet another reason to use caution in reviewing such studies. A technique known as meta-analysis—an analysis of analyses that pools all data from many studies on the same subject—can produce results that apparently contradict many of the individual studies.

Hundreds of studies, for example, concluded that delinquency prevention programs did negligible good. But a meta-analysis by Mark Lipsey showed a small but real positive effect: a 10 percent reduction in juvenile crime. At the same time, he found that "scare 'em straight" programs led to higher delinquency rates compared with those of control groups.

Meta-analysis works, Lipsey explains, by clearing the background "noise"* that comes from doing the research on the real world, instead of in a laboratory. Say researchers are collecting data on teenage attitudes toward crime. A teenager being interviewed could have a bad memory or decide he doesn't trust the interviewer; or the interviewer could have an off day. Even objective measures such as arrest records have statistical noise, Lipsey says. "That may vary from officer to officer. It's not just a function of how the kid does." Some officers in some neighborhoods may simply make more frequent arrests.

Sampling errors are also common, he says. "From the luck of the draw, you get a group of kids that is particularly responsive or resistant. And all those quirks come through in that study." Individual studies, amid this buzz, may not find a statistically significant effect. By pooling data with meta-analysis, however, "the noise begins to

*See Chapter 8, "The Signal in the Haystack."

cancel out," Lipsey says. "Suddenly you begin to see things that were in the studies all along but were drowned out."

Another dramatic reversal in the story numbers tell came in a meta-analysis released in April 1994 on the effect of school funding on pupil performance. Previously, studies suggested that pouring money into teacher salaries and smaller class sizes made a negligible difference. But when Larry Hedges of the University of Chicago reviewed several dozen studies, he found that money made a difference. Meta-analysis, he says, was able to see through the obscuring noise. "People who didn't want to pay for schools used to cite studies showing that funding didn't make any difference," he says. "So these results were very influential."

In the end, a correlation is no more than a hint that a relationship might exist. Without a plausible mechanism—that is, a way that one thing might cause another—it's practically useless. Therefore, it's unlikely that the surge in Wonderbra sales caused the 1994 Republican Congressional election sweep, even though the trends were closely linked in time. On the other hand, studies linking rising teenage obesity to increased hours of TV viewing at least offer a way to get from cause to effect without straining credibility.

Studies that link IQ to race ultimately sink under the absence of a realistic mechanism for linking the two. Evolution is too slow and the differences between races are too muddled and too small to account for the apparent statistical divergence. To do the kinds of experiments necessary to prove the link in humans would be unthinkable, says mathematician William Fleishman of Villanova University. Such research would have to involve random mating and perfectly controlled environments.

Consider a "thought experiment" often used to illustrate this point. Say you found a jar of seeds in your basement and decided to plant them. The jar contained a mixture of seeds, some good, some bad, some fresh, some moldy. You decide to plant half the seeds in your front yard, half in the backyard.

It turns out that your neighbors across the street own a particularly nasty pit bull, so you rarely set foot in your front yard. The backyard, however, is a lovely private haven, so you spend a lot of time there nourishing your seeds, watering them, feeding them, weeding them, making sure they get enough sun, pruning them when they get too tall and begin to fall over.

It won't come as a surprise to anyone to find that the plants in the backyard are thriving while those in the front are going to, well, seed. Of course, some individual plants in the front yard (grown from especially strong seeds) would thrive anyway, due to stamina and luck, and individual plants in the backyard would wither and die, due to old or moldy seeds.

But how would you know which front yard plants were puny due to neglect, or inheritance? How do you know which backyard plants thrived because they were pampered, or smart?

The same is true of children—except that people still don't agree on what kind of nourishment kids really need. "Here we seem to have these highly heritable traits," says Lipsey. "But what is it we know about what's really important to the successful education of children?" Every correlation, he says, should come with an automatic disclaimer. "There's a big logical fallacy here. What you need is a mechanism. But the numbers can be oh so seductive."

Curiously, the very reason that people are prone to jump to conclusions based on tenuous correlations may have something to do with human genetic endowment, according to Paul Smith, who has been analyzing social studies since the early 1970s. "You and I don't have a statistical facility in our brains," says Smith, who as of this writing is at the Children's Defense Fund.

> We are primates evolved to gather fruit in the forest and when possible to reproduce, and I think it's marvelous that we can do what we do.
>
> But we have to exercise almost intolerable discipline to not jump to conclusions. There might be a banana behind that leaf, or it might be the tiger's tail. The one who makes the discrimination best and moves fastest

either gets the banana or gets away from the tiger. So this leaping to conclusions is a good strategy given that the choices are simple and nothing complicated is going on.

But at the level of major social policy choices, [jumping to conclusions] is a serious concern.

In fact, humans as a species are notoriously bad at certain kinds of mathematical reasoning. It's not unusual for people to think they have to invoke psychic powers when only probability is at work. How many people do you have to put into a room to make it more likely than not that two will share a birth date? Answer: two dozen should do nicely. (This seems counterintuitive because we automatically think how many people it would take to match our own birthday; when any matched pair is possible, the probability shoots up sharply.)

The size of your sample can also have a wildly deceptive effect. You might be impressed, for example, if I told you that half the cars on my street were BMWs—until you learned that there are only two cars on my street.

In *Strength in Numbers,* mathematician Sherman Stein offers the case of the men's support group that wanted to demonstrate how badly women treat the male sex. As supporting evidence, the group pointed out that more than half of the women on death row had murdered their husbands, while only a third of the men on death row had murdered their wives. What the group neglected to mention, says Stein, was that there were a total of seven women on death row. And 2,400 men.

Chapter 13

THE BURDEN OF PROOF

PROBABLE TRUTH: A CHANCE IN A MILLION

Strictly speaking it may even be said that nearly all our knowledge is problematical; and in the small number of things which we are able to know with certainty, even in the mathematical sciences themselves, the principle means for ascertaining truth—induction, and analogy—are based on probabilities; so that the entire system of human knowledge is connected with the theory set forth in this essay.

—*Marquis de Laplace, in*
A Philosophical Essay on Probabilities, *1814*

As far as the laws of mathematics refer to reality, they are not certain, and as far as they are certain, they do not refer to reality.

—*Albert Einstein*

There's something about probable truth that seems not quite scientific. "Probable" sounds too much like "probably," which slides all too easily into "perhaps," which in turn can be interpreted as "who knows?"

Yet probable truths are highly quantifiable. They can be calculated very precisely. In fact, in most of the so-called hard sciences, probable truths are the only kinds of truths possible.

Present-day physicists struggle with this problem frequently, trying to explain how something can be accepted as "true" and as merely "probable" at the same time. "In our society, people don't realize that a lot of scientific announcements are probabilistic interpretations of data," says Nobel laureate Burton Richter, director of the Stanford Linear Accelerator Center. "There are some odds that they are wrong."

The idea that the best science can do is give odds on the truth goes back at least as far as Galileo, according to Harvard historian Gerald Holton. "In the historic view, that's where the notion of absolute truth disappears."

It was Galileo who abandoned the more cerebral, theory-based truths and decided that the ultimate arbiter should be experiment. Experiments test whether or not theories are true. Every time the experiment turns out the way theory says it should, the theory gains a degree of validity; it gets more and more true. But it cannot ever become completely "true" unless the scientist performs an infinite number of experiments. That holds the door open for revisions based on new revelations. Laws of nature can be proven untrue, but rarely true. Scientific truths are always provisional.

"Experiment makes a hypothesis more and more probable," says Holton. "But it cannot verify it. That was a significant relaxation of the requirement for useful truth in science. . . . From that time on, truths were statistical."

One of the first people to rely on this method of proof was Ludwig Boltzmann, who applied it to the relationship between disorder and irreversibility—that is, the more scrambled things get, the less likely that they will ever come unscrambled. Or put another way, the more disordered things get, the more energy it takes to get them ordered again.*

Boltzmann saw that he could not "prove" his hypothesis true in the normal sense. But he argued, in 1877, that it was true in the overwhelming majority of cases, hence with a very high probability. "This remarkable statement," write the authors of *The Empire of Chance*, "implied for the first time that a law of nature may not hold with necessity but only with probability, and hence allow for exceptions."

In fact, probability permeates just about any attempt to pin down a scientific "fact." The fact itself might wiggle away from any precise attempt at measurement, or the measurement (or measurer) might be wobbly or overwhelmed with background static and interference.

Take a straightforward "fact" such as my height. Recently asked how tall I was in a doctor's office, I answered five feet five inches. Then I added: "At least in the morning."

The sad truth is, by evening, I've shrunk at least an inch. And so, dear reader, have you. There's no mistaking that gravity gets us down—everybody equally. Give it a full day to pull us toward the center of the Earth, and we compress like an accordion. (If you don't believe me, just try taking your height when you first wake up and then again before you go to bed.)

This natural variability means it makes little sense to measure height to an accuracy of a millimeter. Gravity aside, height changes every time you take a breath, or shift your weight, or think tall thoughts. In that context, a difference of a millimeter would have no meaning.†

*See Chapter 6, "Emerging Properties: More Is Different."
†See Chapter 4, "The Measure of Man, Woman, and Thing."

"Every measurement is approximate in a certain sense," says physicist Haim Harari, who is president of the Weizman Institute of Science in Israel. "There is no point in measuring the distance from New York to L.A. to the last inch, because it depends on whether I measure it into your kitchen or your bedroom."

The other thing that varies is the experimental apparatus, including the behavior of the experimenter (if we consider them both of a piece). When I measure my daughter's height, I push the book down very firmly on her head and draw the line under it with a thick pencil, then measure the distance from the bottom of the line to the floor. (In our household, height is highly prized and only the cat is shorter than me, so I like to get every advantage I can. To some extent, this kind of wishful thinking enters into all measurements.)

Even if I were perfectly objective, however, and even if her height were always the same to the last millimeter, the measurement would still have lots of wiggle room. The ruler would expand if the room was warm, contract when cold. A slight angle in the placement could make a measurable difference. No hand is perfectly steady. It wobbles because it's alive—by reason of the pulse alone.

So what could I honestly answer to the supposedly "factual" question: How much has your daughter grown this year?

If I measure an inch, does it mean she grew an inch? Is that a fact?

Or have I just measured the effects of gravity, or room temperature, or posture, or even mood (is she feeling low today?), or a particular tilt of head? Or maybe I stretched the measuring tape, or started it at a different place, or unconsciously inched it upward.

How can I know?

Probability is tailor-made to sort this kind of scenario out. First you calculate the probability that the measured inch was due to experimental error. Then you calculate the probability that the measured effect was the result of natural variations (like gravity or posture). If you subtract both of those, and you still have a measurable effect (say, an inch), then you know the result is real. (She grew again!)

Most everything in physics has to account for the probability of error in some way. Science historian Ted Porter of UCLA puts it this way: "You don't see a muon [a species of subatomic particle] jumping out at you. You see a track and you try to infer how likely it is that it's a muon. But you also have to factor in: What kind of apparatus do you have? What kinds of people are doing the work? Are they trustworthy? How was the detector working that day?" (This might put into perspective the reasons O.J.'s defense lawyers kept hammering on the reliability of the technicians who prepared the DNA samples and performed the DNA matching.)

And that's not the end of probability's pervasive role in seeking out "truth." After you account for every conceivable natural variability, every conceivable experimental error or prejudice, every bit of orneriness of every machine or measuring device, you still have to factor in fate.

What are the odds that the "fact" you've found is simply a fluke? A mere coincidence? A truly random event?

Particle physicist Leon Lederman ran into such a quirk of fate when he first "discovered" the "upsilon" particle—which came to be known as the "Oops, Leon."* "You try to be as numerical as you can," says Lederman. "Take one of my disasters." Lederman recounts how his experiment found a telltale "bump" in a curve that appeared to signal the presence of the upsilon particle. Particle physicists are known as "bump hunters" because they plot data and look for sharp peaks where something unusual seems to be happening. If you made a graph of random coin tosses and plotted the numbers of heads and tails, you would get just such a "bump" if you happened to turn up, say, ten heads in a row.

*See Chapter 8, "The Signal in the Haystack."

"First you ask, what's the probability that events will cluster there just by random pileup?" Lederman continues. "That's a fairly straightforward mathematical operation. It turned out it was one in fifty." In the case of the upsilon, "we found this bump, with one chance in fifty of it being chance. It turned out to be chance. Oops!"

The confusion over cause and coincidence is one reason that probability kept popping up in the O.J. trial. If DNA tests showed that the blood on O.J.'s Bronco matched Nicole Simpson's blood, what is the probability that the test results are due to pure coincidence?

Individual DNA "signatures" appear as rows of dark lines spaced out in a pattern that looks a lot like supermarket bar codes. The dark lines appear where similar fragments of DNA line up. (The process is not so different from throwing a bunch of sand, oil, and water into a jar and mixing it up; as gravity pulls on the ingredients, they'll tend to line up in layers, with the heaviest at the bottom.) Since each person's DNA is unique, each pattern of lines spells out a unique signature. So if two patterns match, you can bet that the DNA came from the same person. At least most of the time. You still have to factor in the very real probability that the match is due to chance. What those probabilities are is a matter of much ongoing debate. (Clearly, if people shared DNA profiles at the same rate they share birthdays, they wouldn't be much use as evidence in court—simply because coincidence plays too great a role in any "positive" result.)

The question of coincidence also came up when Joseph Fraunhofer discovered that starlight contains a kind of bar code that can be read as the "signature" of elements on the star—just as DNA mapping can be read as genetic "signatures" of people.* But before astronomers could accept the dark lines as fact rather than quirk, they had to calculate the probability that the dark lines in the starlight were simply random streaks.

Fraunhofer first noticed the lines in the spectrum of our own

*See Chapter 4, "The Measure of Man, Woman, and Thing."

neighborhood star, the Sun. Spread out in a long band of rainbow colors, the Sun's spectrum is interrupted with black lines like black keys on a piano. Similar patterns of *bright* lines are seen in the spectrum emitted by every chemical element. Heat up sodium or neon or iron or helium until it glows, and a unique pattern of brightly colored lines appears as atoms spit out precisely measured chunks of light. The spectra are the fingerprints of atoms, as unique as a DNA signature.

So Fraunhofer wondered: The lines of color emitted from the elements seemed to line up exactly with the dark lines in the spectrum of the Sun. Did that mean the Sun was somehow absorbing parcels of light that corresponded to specific elements? If so, that meant that one could read out the ingredients of a star simply by looking at the spectrum of dark lines, the stellar DNA.

But first Fraunhofer had to rule out coincidence. He figured out that the probability that any line in, say, the iron spectrum would fall in the same place as a dark line in the Sun's spectrum merely by chance was 0.5. But the odds that all sixty lines emitted by glowing hot iron would match up with the shadows in sunlight would be 0.5 multiplied by itself 60 times (or raised to the 60 power). That was an incredibly small number—easily small enough to rule out any significant role for chance.

Today, astronomers routinely use Fraunhofer lines to "see" the chemistry of distant stars.

A recent example of statistical truth-seeking was the "discovery" of a subatomic particle called the top quark at Fermi National Accelerator Laboratory. Quarks are produced in collisions of subatomic particles at enormous energies. Unseeable themselves, they are ultimately identified by the tracks* they leave in particle detectors.

But hunting a quark is like looking at a snowy trail packed with

*Actually, quarks don't stick around long enough to leave tracks; they condense into other kinds of particles in a subatomic eye blink. Those secondary particles leave tracks, which in turn become the evidence for the quark's fleeting presence.

hoofprints to find the print of the particular animal you seek. The individual prints are hard to see, and many look alike. What seems to have the clear marks of the quark might be some entirely different species with similar prints (a red herring thrown in by nature, for example).

"All we can do is measure the probability that certain attributes were produced by the quarks and a certain probability that those attributes were produced by more prosaic processes," says William Carrithers Jr., head of one of the teams that found the quark. "The whole discovery is an inherently statistical process."

Physicist Nicholas Hadley of the University of Maryland, a member of another quark-hunting team, made this process clearer with a coin-toss analogy. Assume a physicist wants to find out whether a coin is the normal heads-and-tails variety or an unusual species with two heads. Then assume that the scientist can't look at both sides directly but can only toss the coin and see what face comes up. How do you distinguish between a normal coin, with a heads and a tails, and a coin that has two heads?

Getting three heads in a row wouldn't be convincing evidence that the coin had two heads. That happens one out of eight times with a normal coin. "But if you do it a million times and get only heads, then you know the coin has two heads," says Hadley. "By being careful and clever, you can arrange it so it takes fewer and fewer 'tosses' to say that it has to be the top quark."

One way of being clever is to realize that flipping one coin a million times and getting a million people to flip a coin once are equivalent, probability-wise. Let's say the probability of getting ten heads in a row is one in a thousand. Then if a thousand people flip coins, it would be normal for one person to get heads ten times in a row. The experimenters searching for the top quark arrange their experiments to conduct such "flipping" billions and billions of times, creating conditions tailor-made for the top quark to make an appearance.

The innate uncertainty associated with probability makes DNA profiles a tricky way to "prove" guilt—especially to jurors unfamiliar with the perils of statistics. When short strings of DNA from a spot of blood or semen found at a crime scene are compared to similar strings of DNA from a suspect, the odds that another individual could share that same profile are usually given as less than one in a million. But in a country of 250 million people, that means 250 people could share the same genetic profile.

How good, then, do the odds have to be to consider something to be "true"?

The answer is, it all depends.

In the case of the top quark, physicists made an initial announcement that they had found "evidence for" the quark when there was a 1 in 400 chance they were wrong—about the same chance as pulling two aces from the top of a deck of cards. Before they went public with an official announcement that they had discovered the top quark, however, they waited until further experiments produced a certainty more like 1 in 10,000—that is, a 1 in 10,000 probability that the results were due to chance.

The degree of tolerance for chancy results also depends on whether there's a strong prejudice in favor of the result. If the physicists had seen the Energizer Bunny pop up in their atom smashers, then they would certainly have demanded a more stringent set of requirements than for the top quark—which was not only expected, but practically required by the currently accepted theory. In fact, if the quark hadn't been found, the generally accepted model of how matter is constructed would have been all but destroyed.

The odds against a mistake in DNA matching have to be much greater because a human life, not just a scientific discovery, is at stake, and scientists have battled about precisely what those odds should be. In part, the strength of the odds has to do with the number of times the coin is tossed—or in this case, with how many stretches of DNA from a suspect are compared with DNA from the scene of the crime.

The more stretches that match, the lower the probability that any particular match is due to random chance.

If the DNA samples have been contaminated (as was charged by the Simpson defense), all these results are worthless. But even bad DNA evidence, many biologists and legal experts say, is more reliable than other kinds of evidence that juries hear.

In fact, the battle over DNA matching has put all courtroom evidence on trial, at least in the eyes of many scientists. Steven Austad, a biologist at the University of Idaho, cited such dubious "evidence" as fingerprints, ballistics, and the famous barking dog of the O.J. trial. Worst of all is eyewitness testimony. "Any psychologist could tell you that kind of identification is prone to error," he says. "Misidentifications are rife. There's a large psychological literature on this."

People demand much higher standards of proof for DNA evidence than for other kinds of evidence. "People worry, gee, there's a probability of 1 out of 10,000 that there could be an error," says Austad. "Yet the things they consider to be the gold standard aren't even close [to 1 in 10,000]." Evidence that rates as gold standard among jurors includes eyewitness testimony, often the least reliable of all. Human memory is notoriously fallible and prone to alteration over time.

The relative reliability of DNA evidence has turned out to be a great tool for freeing people wrongly convicted on the basis of fingerprints and eyewitness testimony—and later proved innocent by DNA matching. The Innocence Project at the Benjamin Cordoza School of Law in New York, for example, has overturned several dozen convictions using DNA matching. All were people convicted on eyewitness testimony and circumstantial evidence, but exonerated by DNA matching.

Although a match between DNA profiles of the suspect and a spot of blood at the crime scene could be due to coincidence (although the probability might be small), the absence of a match is positive proof of innocence. "With matches, you get a probability," says Jonathan Oberman of the Innocence Project, "but exclusions are exclusions."

The same reasoning applies in physics: It's possible to say for certain that a particular particle track cannot be the top quark. But it's only very, very, very probable that a particular track is the top quark.

In the end, the degree of confidence we place in probable truth depends largely on the context. "Remember the old game called truth or consequences," says Richter. "If the consequences of the mistake are that somebody is going to prison for a long time, or be executed, the public demands an extremely high standard of truth, and that's proper. If the Fermilab people say we've found the top quark and we're sure to one in a thousand, and they're wrong, no one has been put in prison. And the Nobel Prize committee will wait long enough to find out."

LOGICAL TRUTH: IT STANDS (OR FALLS) TO REASON

Mathematics may be defined as the subject in which
we never know what we are talking about,
or whether what we are saying is true.

—*Bertrand Russell*

The wonderful thing about mathematics is, it couldn't care less about subject matter. Two apples plus two apples is four apples the same as two planets plus two planets equals four planets, or two ideas plus two ideas equals four ideas. It's so impersonal that most of the time it doesn't even use recognizable labels. $2x + 2x = 4x$ will do just fine.

Math has its own inherent logic, its own internal truth. Its beauty lies in its ability to distill the essence of truth without the messy interference of the real world. It's clean, neat, above it all. It lives in an ideal universe built on the geometer's perfect circles and polygons, the number theorist's perfect sets. It matters not that these objects don't exist in the real world. They are articles of faith.

No wonder mathematics has sparked an almost religious fervor among its practitioners. Like religion, mathematics has a set of clear rules for telling right from wrong. Step by step, you follow a chain of logical arguments to the truth; if you stay on the path, you can get from here to there without a misstep.

Getting to the truth in the real world requires dealing with probabilities, natural variations, the difficulties of making measurements. But mathematics is immune from messiness and ambiguity because it builds on logic alone. The whole thing is constructed from perfect blocks of logical propositions. Its hands are clean, its process unsullied. Like George Washington, numbers never tell a lie! (Or so we like to think.)

Still, we sometimes draw the wrong conclusions from even the simplest logical statements. For example, one logical misunderstanding came up again and again in the O.J. trial. Many people took the statement "If the DNA matches, he's guilty" to be equivalent to the statement: "If the DNA doesn't match, he's not guilty." But while both may seem to say the same thing, the first conclusion is false, while the second is true. Proving that a DNA sample cannot possibly come from a suspect is a straightforward matter; if it doesn't match, he goes free. But a match between the DNA of the suspect and the DNA from blood at the crime scene doesn't absolutely prove guilt. In only indicates very, very, very, very probable guilt.

Or put more simply, if Johnny hates vegetables and peas are vegetables then it can be proved without a doubt that Johnny hates peas. But the reverse is not necessarily true: If Johnny hates peas, you cannot conclude that therefore he hates vegetables.

Until the 1930s, it was generally assumed that any mathematical statement could be proven true or false by logic. But then a mathematician named Kurt Gödel proved that there were truths beyond logic, truths that logic alone could not prove. This was a shattering blow. "Formal deduction has as its crowning achievement proved its own incapacity to make certain formal deductions," writes James R. Newman in his *World of Mathematics.*

In essence, Gödel's notorious theorem says that some statements in mathematics cannot be proven true (or false) within their own logical universe. At some point, it always becomes necessary to jump out of the local universe in order to prove that a statement is true. In a way, this is like trying to prove that two islands are connected underneath when you're stuck on one of them. It would be much easier if you could somehow jump off the ground and get a bird's-eye view—from a plane, say. Gödel proved that in a mathematical universe, this jumping out of the system is sometimes required.

Take the statement "This sentence is not true." If it's true, then it's false. But if it's false, is it true? One goes around in circles without getting anywhere. There is no consistent or logical way to prove the truth—or falsity—of the statement, primarily because the statement is talking about itself. It's stuck in its own logical system and can't get out, so it can't prove the truth or falsity of what it says.

Or take the sentence "I am lying." True or false? If you say true, and I am really lying, then I'm telling the truth about lying, so I'm not lying. On the other hand, if you say false, I'm not lying, then I must be lying because the statement is true.

Mathematical truth was supposed to be irrefutable because it was based on logic. But paradoxes like this revealed that logic can lead to contradictory conclusions. Logic cannot be a clear guide to truth if it can point in both directions at once, or nowhere, or to nonsense.

This notion that a logical statement could point in two different directions violated one of the most cherished commandments of logic—Aristotle's law of the excluded middle. According to Aristotle, there was no middle ground and no way that logic could lead to contradiction. A pea was a vegetable, or it was not. A statement was true, or it was false. Things were black, or white. When Hamlet agonized: "To be or not to be!" or Patrick Henry declared: "Give me liberty or give me death!" there was no fudging around. One was one, or the other. (Funny how inadequate Aristotle seems today: If the meaning of life, death, and freedom were clear-cut there would be no national debate about abortion or assisted suicide. Sometimes it seems that middle ground is all there is.)

A hunger for that lost Aristotelian clarity continues to gnaw at us today; how much simpler it is to live in a world of sharp edges and absolutes, where directions and doorways are clearly marked on the physical and philosophical map.

This nostalgia for the excluded middle leads us to all kinds of nutty conclusions. "Human beings love to divide the world and its inhabitants into pairs of opposites," writes Carol Tavris in her book *The Mismeasure of Woman*—male and female foremost among them. But the minute you start defining male as not female, and vice versa, we're left with a world in which real men don't eat quiche—or do a multitude of womanish things they might otherwise naturally do, and a world in which women are similarly restricted to activities defined as "not male."

Tavris offers up a very old lesson in logic taught in all introductory texts. The reasoning goes: All men are mortal. Socrates is a man. Therefore, Socrates is mortal.

Now try it this way: All men are mortal. Alice is . . . ?

Aristotle was trying to show that you can't contradict yourself and still be logical. "The same thing cannot at the same time both belong and not belong to the same object and in the same respect," he wrote. "This is the most certain of all principles." Yet we do it all the time.

The rule of the excluded middle has even been codified into computer language, and in some cases, found wanting. A computer program doesn't naturally deal with shades of gray; things are black and white, yes and no, on and off. It is a binary world of ones and zeros.

Some years ago, a professor at the University of California, Berkeley, decided there was a better way. After all, he reasoned, very few real-life phenomena are easily sliced into clean categories or divided along clear lines.

What is poetry? Keats? A ditty that a computer turns out? What is a chair? Is it a stump? A dollhouse chair? A dog bed? Where does the chair end and the air around it begin? What is hot, warm, cold, cool? What is dead and alive? Who is handicapped, or crazy? What is a planet, or a star?*

So the professor, Lotfi Zadeh, proposed and then developed a field known as "fuzzy" logic.† Fuzzy logic allows for a spectrum of possibilities. It assigns values. Instead of having to choose between black and white, it gives you a sliding scale of values, say, from 1 to 10. You can even be fuzzy about fuzziness. If you can't decide whether it's an 8 or a 9, give it a 9.2.

Fuzzy logic goes way beyond the true/false requirements of two-valued logic. With two-valued logic like Aristotle's, it will rain tomorrow, or it won't; a person is employed, or not. With fuzzy logic, a person can work part-time, and the weather can drizzle, uncertain of its own intentions.

Fuzzy logic caught on big in Japan and, according to Daniel McNeill and Paul Freiberger in their book *Fuzzy Logic,* has been applied with enormous success to all kinds of technology, including household appliances and superfast trains. It has met with less success in this country, McNeill and Freiberger argue, because "fuzzy thinking" is associated with garbled thinking. Yet, it has made some inroads. NASA has adopted fuzzy logic as part of its strategy to control dockings between the shuttle and international space station and to guide rovers on the surface of the Moon or Mars.

It's a great gimmick for feedback devices like thermostats, where "hot" and "cold" don't mean much out of context. "Much, much

*See Chapter 4, "The Measure of Man, Woman, and Thing."

†Like complexity theory, the ideas behind fuzzy logic had been in use for some time before the idea acquired its sexy name.

cooler" or "hotter than I like it" are much more appropriate and much more amenable to quantification by fuzzy logic.*

The resurrection of the excluded middle and the migraine-inducing deduction of Gödel dealt a staggering blow to our sense of certainty—especially about the truth to be found in numbers. "Not even mathematics—after Gödel we find doubt there too—can give full surety," writes physicist Philip Morrison.

The shock was felt far beyond mathematics. In 1988 the *Los Angeles Times,* no less, was so taken by an extension of Gödel's work that it commented on it in an editorial. The new result, the editors concluded, "makes the world shake just a little."

Challenges to common sense are certainly not new in science. We have come to accept such absurdities as the roundness of Earth (with half of the world's population dangling upside down) as a matter of course. We accept that the selfsame atom—simple carbon—makes diamonds, coal, pencils, and (mostly) us.

As Newman points out, people even seem comfortable with the fact that there is no distinction between matter and energy anymore—or that although only a short time ago we believed ourselves descended from the gods, we now visit the zoo with the same friendly interest with which we call on distant relatives.

Yet we cannot accept absurdity in mathematics. Mathematics is supposed to lead us not into absurdity, but to certainty. The long chains of reasoning are supposed to get you from here to there without a misstep or wrong turn. To assert otherwise seems blasphemy.

Mathematician Morris Kline laments, in his book *Mathematics: The Loss of Certainty:* "This confidence that truths would be discovered in all fields was shattered by the recognition that there's no truth in mathematics."

*Alas, fuzzy logic does not address the problem of logical contradictions or paradoxes.

Logic is only a sign that points "That way!" or "Look here!" It may be a hoax. But the only way to tell is to follow it and find out.

What Gödel boils down to, says Kline, is that "the price of consistency is incompleteness." Mathematical truths are no more complete and all-encompassing than other kinds. Neither are they any more "self-evident." Those of us reared on Euclid swallowed without thinking all those axioms about the obviousness of such propositions as: Two parallel lines never meet. Yet one only needs to look at the lines of longitude—which are parallel at the equator—to see that they do. The obvious becomes clearly false and the false clearly obvious.

"Obviousness is always the enemy to correctness," writes Bertrand Russell. Logic is a useful tool, but it, too, has its limits.

Indeed, the limitations of logic, some mathematicians say, may be the main reason we don't yet have fully intelligent computers, like Hal in Arthur Clarke's *2001*. Traditional approaches to artificial intelligence assumed that thought could be programmed into computers in part because thought itself was logical. But increasingly, evidence suggests that the road to intelligence is not reason—or, at least not logical rules.

In his book *Goodbye, Descartes,* logician Keith Devlin argues that the logical road has led to a dead end. "Descartes argued that the rational approaches provided the 'right' way to look at the activity of the human mind," he writes. However, a logical approach requires—among other things—that "words have fixed, definite and unique meanings. But that is just not true."

Consider the way the brain tells the nouns from the verbs in the sentence "Time flies like an arrow." Now consider the very different way it interprets the same words in the sentence "Fruit flies like a banana." No rule-based artificial intelligence system could ever make that leap, he argues. Devlin's working on a new kind of logic—which he calls an "algebra of conversation"—that might come closer to capturing what really goes on within and between human minds.

This new logic would be "soft mathematics," a departure from the precise, rigorous forms of the past. Traditional logic works in any context at all—that's what gives it power. $x + x = 2x$ no matter what x, y, and z are. However, soft mathematics would take context into account. And it's the brain's ability to consider context that keeps it from tripping over "flies" or "like" as it flips between two very different meanings in the sentences above.

In the same way, the reason "this sentence is false" is a paradox is because we read it in two different contexts: the context of what the sentence is saying, and the context of what the sentence is saying about itself. Once you take the context into account, Devlin says, the statement is no more paradoxical than the "conflict between an American who thinks that June is a summer month and the Australian who thinks June is a winter month."

Soft mathematics will require metaphor as well as formal mathematical reasoning. It may not even be mathematics, Devlin says. But it will be necessary to take the next step. To really understand what it means to think rationally, mathematics will need to team up with psychology and sociology, and perhaps even biology and poetry.

All this means is only that mathematics is, like all of us, perfect and flawed at the same time. Once upon a time, people believed that the laws of logical deduction—and the laws of nature they explained and described—were set in stone. Now mathematicians, even, know that there are many different kinds of logical systems, that logic itself has limitations, and that laws of logic and nature evolve just like legal laws change (or should) to reflect changes in society.

People often think legal concepts should be immutable because, after all, the highest laws—the laws of nature—are immutable. Ideas such as the right to bear arms or majority rule solidify into dogma and even seem to derive a certain justification from a belief that they are following nature's way—or the way of "science."

But natural laws are rarely dogma. They work only within well-defined limits. Newton's laws of motion and gravity will get you to the Moon, but they don't work in extreme situations, for example, when you are traveling almost as fast as light, or when the pull of gravity is enormous. Einstein's relativity is much more appropriate to these contexts and leads to conclusions Newton's laws could never have predicted, like the existence of black holes.

When the laws find themselves operating unexpectedly in a new context, there is no earthly reason the rules should stay the same. It should not be surprising, therefore, that simple laws of logic do not apply in the complex landscape of the human brain. Parallel lines do meet in curved space, even on Earth.

TRUTH BY CONSENSUS: A JURY OF ONE'S PEERS

Since neither science nor law seems to have a foolproof method of getting at the truth, they must find a way to come to a working consensus while awaiting such definitive evidence as a confession or the ability to conduct more sensitive or more imaginative experiments. Science has the easier task, because sooner or later, nature will show her hand and set things straight. But law can't wait for the luxury of complete information; it has to decide.

The Supreme Court recently put it this way: "Scientific truths are subject to perpetual revision. Law, on the other hand, must resolve disputes finally and quickly."

There are times, however, when the different standards of proof in law and science find themselves on a collision course. The validity of scientific evidence presented in a court is certainly one of them. This is happening more and more often. In fact, scientific experts are becoming so common in courtrooms, says Paul F. Rothstein, of

Georgetown University, that it's a rare piece of litigation that doesn't bring in a scientific expert at some point in the proceedings.

Yet because of the severe consequences of a wrong call, science in court is held to a higher standard of proof than it is among scientists. "The courts expect standards of demonstration that are impossible in science," says UCLA science historian Ted Porter. "They expect scientists to behave like reasoning machines. But if they behaved like reasoning machines, they wouldn't get anywhere."

Science and law also collide when it comes to assessing risks, benefits, and blame for the consequences of new technologies. As Hal Lewis points out in *Technological Risk,* the legal system behaves as if causes can always be found for misfortunes associated with technology—if only enough money and effort and experts are put on the case. But, he concludes, "it isn't always so."

When the National Transportation Safety Board investigates the cause of an airplane or train accident, it doesn't finish its investigation by assigning cause—only "probable cause." And even then the relationships between cause and effect are uncertain.* As Lewis points out, "The use of cause-effect relationship in risk analysis involves uncertainty in both directions." That is, a cause may not have a single, easily definable effect (medicines may create symptoms, as well as ease them). Also, an effect may have many different causes—say, a gene that causes breast cancer may not pose a danger unless other genes or environmental factors are brought into play.

"Uncertainty is not a dirty word," writes Lewis. "Every scientific measurement and every scientific estimate involves some degree of uncertainty, sometimes small and sometimes large, but always present."

It would be nice if some of that respect for uncertainty could find its way into the law, and in some sense it does. For example, English

*See Probable Causes in Chapter 12, "The Truth about Why Things Happen."

law, according to John Barrow in *Pi in the Sky,* has its own version of fuzzy logic. Unlike O.J., who could only be judged "guilty" or "not guilty," a criminal defendant in Scotland can be judged "not proven" guilty, which leaves open the possibility of a second trial. This is much closer to the way science goes about getting to the truth.

Since there's normally no way for the truth to be absolutely decided beyond doubt, both law and science place their trust in a process called "peer review"—or trial by a jury of peers. What it means is simply that a group of people much like yourself (peers) sit in judgment and come to a consensus about the truth of the matter in question.

Lawyers take their cases to juries, and scientists submit their ideas in the form of papers to the editors of scientific journals, who send them out for evaluation by fellow scientists. While peer review generally works over the long haul, it doesn't always work in the time frame relevant to courts. In some cases, it has been notorious for being wrong, especially about cutting-edge science. Science historian Gerald Holton of Harvard likes to point out that Einstein's seminal paper on relativity almost didn't get published because there was only one person in all of Germany who would believe it. "A lot of correct and even great science did not pass peer review," says Holton.

Still, until recently, peer review of scientific ideas stood as the standard for expert testimony in court. That is, scientific evidence could only be presented in court if the method or science in question had already passed "peer review" of the scientific community itself.

Then, in a case against Merrill Dow Pharmaceuticals by a family who said their child had been harmed by the morning sickness drug Bendectin, the Supreme Court struck down the peer-review standard. The family wanted to present in court scientific data that had not been published in peer-reviewed journals. While the family ultimately lost their case against Dow, the court decided peer review was far from a foolproof method for evaluating good science.

Even when formal peer review is not involved, science depends mightily on consensus. Scientists, like lawyers, have to persuade their colleagues that their results are believable, uncontaminated, and perhaps even right. Sometimes the consensus is temporary. (Witch trials, like the widespread belief in humors and ethers, were based on such temporary consensus. Consensus doesn't mean validity, just agreement.)

The top quark is (once again) a case in point. Back in 1985, its discovery at a European laboratory was proclaimed on the front page of papers around the world. A U.S. physicist, reading of the discovery, remembers being highly amused. After all, he says, the detector that found the quark had been shut down months ago for repairs, and since that time scientists had been arguing about how to interpret the results. The announcement, he says, "only means that the physicists finally agreed on what they saw."

They didn't agree for long, and soon afterward the "discovery" was retracted.

Almost ten years later, the quark was discovered again at Fermi National Accelerator Laboratory near Chicago. This time, the discovery was on much firmer ground. Yet even so, the top quark was not officially unveiled at a press conference until after months of negotiation to get some nine hundred physicists to sign off on the final wording of the papers.

When the Supreme Court struck down peer review as the test for scientific validity, they replaced it with a very unscientific-sounding concept: judgment. That is, they threw the responsibility for evaluating the scientific "truth" back to courtroom judges, who are now charged with evaluating a scientist's credentials and methodology themselves.

Deciding what science to trust, and what to throw out as shoddy, is often based on judgment in science as well. A researcher's peers may look as hard at the reliability and reputation of the experimenter as they do at the experimental results. When word leaked out that NASA researchers had found evidence of ancient fossils on a meteorite from Mars, the only reason many scientists took the results at all seriously was that a well-respected Stanford chemist—Richard Zare—was a member of the team.

Trusting the expert judgment of other researchers, says Porter, is both necessary and common in science. "It's what people commonly call expertise." People who spend their lives studying stars, or genes, or baseball are in a better position to make sound judgments about events in these areas than people who do not.

Judgment played a central role recently in one of the most dramatic mathematical announcements of the century. In the summer of 1993, popular and professional publications alike leaped into a near frenzy at the news that mathematician Andrew Wiles of Princeton had finally solved the famous problem known for three hundred years as Fermat's last theorem.* Since Wiles's proof was highly technical—and ran some two hundred pages—very few mathematicians were qualified to serve as peer reviewers. In fact, at first Wiles showed it only to a handful of people.

"In the past, in principle, you could always go and check out a proof for yourself," says mathematician Keith Devlin, who is dean of science at Saint Mary's College of California. "But the complexity and amount of information [in Wiles's proof, but also in mathematics in general] is now such that we have to rely on other people."

*As most people learned in high school, you can calculate the long end of a right triangle by adding the squares of the shorter sides, then taking the square root of the result. That's because you can solve the equation $x^2 + y^2 = z^2$. Sometimes you can find a situation where the x, y, and z are all whole numbers. For example, $3^2 + 4^2 = 5^2$. (Or $9 + 16 = 25$.) Fermat wrote in the margin of a book that he had managed to prove that no whole number solutions were possible with powers higher than squares. For example, $x^3 + y^3 = z^3$ has no whole number solution. However, he wrote, the margin of the book was too narrow to contain the proof. As simple as this problem may sound, it took three hundred years to come up with a proof. (Indeed, most mathematicians doubt that Fermat really had the ironclad proof he thought he did.)

Eventually, Wiles's proof received a final stamp of "truth" from the mathematical community. But it took almost two years to do. And the final version of the proof that was accepted was quite a bit different from the one Wiles first announced. Glitches appeared and were worked through. One of them took a year to correct. Most mathematicians are convinced that he did, indeed, solve this ancient problem, even though few have read—or understood—the proof.

Confidence in the top quark results also relied on trust to some extent. Physicists believed them in part because they trusted the logic of the experiments and the reputations of the experimenters. The more complex and specialized the work, the more nonspecialists are forced to rely on the expertise of people who do it. "Ideally, physicists establish truth by experiment, and mathematicians establish truth by rigorous proof," says Devlin. "But in truth, we're at everybody else's mercy."

Complicating the issue has been the invasion of the ubiquitous computer into the centuries-old methodology. "It's not just that computers have helped mathematicians to do things they used to do," says Devlin, "it's that they do things differently. The nature of how we prove things has been fundamentally changed by technology." One big change, for example, is in the notion that a proof produces not only "correctness," but also "understanding." What, in other words, does it mean "to know" that something is true?

The first widely accepted computer proof solved a long-standing problem called the four-color theorem—that is, is it possible to prove through logic alone that any map can be colored in just four colors so that no two areas of the same color border each other? The computer proved that the theorem was true—basically by examining a large number of possible configurations of "countries" that were representative of all possible maps. Many mathematicians didn't accept this as a proper proof because it was just a calculation. The computer proved the theorem was true, but not *why* it was true.

Or as Michael de Villiers points out in the scientific journal *Pythagoras,* proof is not about making sure that something is true, but about exposing "the underlying logical relationships between statements." In other words: "Proof is not a question of making sure, but a question of explaining why."

In the end, neither legal experts nor scientists seem clear about the ability of judges to decide what's good science and what's not. "The legal system doesn't seem to be set up to deal with scientific thinking," says Steven Austad, a biologist at the University of Idaho. "What we would really like to know is relative probability of truth in all [kinds of evidence.]"

Yet even if law used perfect scientific reasoning, it still might not find the ultimate "truth"—because not even scientists claim that there is one right answer for every question. "The [legal] briefs seem to assume that if judges simply adopted 'scientific principles' in . . . evaluating expert testimony, the 'correct' result would become apparent," says attorney Joan Bertin. "But the very application of accepted scientific principles often leads scientists themselves in different directions."

What to do?

"One does the best one can," says science historian Lorraine Daston of the University of Chicago. Or as Newman puts it, "The most secure part of our present knowledge is knowing what it is that we do not know."

Chapter 14

EMMY AND ALBERT: THE UNVARYING NATURE OF TRUTH

TRUTH AND BEAUTY

Pure mathematics is, in its way, the poetry of logical ideas.

—*Albert Einstein, in reference to the work of mathematician Emmy Noether, on the occasion of her death,* New York Times, *May 1, 1935*

Albert Einstein knew full well that his theory of relativity* had a philosophical power that went far beyond physics. At one point, he was even moved to remark that more clergymen seemed to be interested in the implications of relativity than physicists.

But if ever a scientific idea got garbled in the translation from equations to popular culture, this is it. Somehow relativity managed to slip into conventional wisdom with a meaning opposite to its true message. While Einstein's theories focused on invariant properties of

*The special theory of relativity deals with the properties of light, matter, and energy, and explains bizarre effects such as time dilation; the general theory of relativity deals with the four-dimensional nature of space-time, which produces gravity and also such bizarre effects as black holes.

things that do not change, no matter what, the popular translation came out as some version of "There is no truth; truth depends on how you look at it"; or simply, "Everything is relative."

Relativity was the fruit of Einstein's search for the unvarying, deep truths that link seemingly disparate ideas together. Deep truths do not depend on fickle appearances. They are the hidden scaffolding that holds the many-faceted house of nature up. Instead of "everything is relative," relativity says: "Things look relative, but don't let that fool you."

Most people have heard of at least some aspects of relativity—for example, the fact that space and time become disturbingly elastic if you're moving close to the speed of light. Certainly, science fiction writers have made good use of this idea—slowing down time, for example, so that astronauts can be gone for centuries, yet return to Earth only a few years older than when they left.

But effects such as time dilation turn out to be comparatively trivial consequences of the surprising fact that the speed of light never changes, no matter what.

Looking for invariants is largely a mathematical exercise, and Einstein worked very closely with mathematicians who specialized in the subject. In the language of math, invariants are usually referred to as symmetries. These are not limited to the same kinds of symmetries we admire in snowflakes and butterfly wings. Something has symmetry to the extent that it is invariant to certain kinds of change. For example, a circle is perfectly symmetrical because it is invariant to any rotation about its center. A square is less symmetrical because only rotations through right angles leave it unchanged. The letter *A* is symmetrical under reflection because its mirror image is unchanged from the original.

The search for symmetry turns out to be a very effective tool for looking beneath superficial differences that camouflage similarities to find a more substantive, permanent meaning. Symmetry therefore lends a satisfying concreteness to the vague sense that there is beauty in truth, and truth to beauty. So many of the things people admire are symmetrical: whether they are natural symmetries, like snail shells, or human-made symmetries, like codes of law that attempt to impose equal outcomes on both sides of an argument. It is nice to know that there's a real quantitative connection between things we admire for aesthetic reasons and things that steer us toward a deep understanding of nature, including, perhaps, human nature.

Scientists have known about this connection for a long time. Physicist Hermann Weyl, who wrote the classic book on symmetry, puts it this way: "My work always tried to unite the true with the beautiful; but when I had to choose one or the other, I usually chose the beautiful."

Beauty in the mathematical sense is a lot more than a pretty face. It is a way of distilling the essence of things out of the messy mix that nature presents us. Edward Rothstein, trained both as a musician and a mathematician, writes that when we search for symmetries, we are "defining which aspects . . . we find essential and which aspects are irrelevant."

In a sense, people make use of the idea of symmetry when they turn ideas over in their minds, seeking kernels of constant truth. You look at the problem this way and that, you turn it inside out, you try to peel off the misleading outer layers to see what's left—what doesn't change as you look at it every way you can.

"What better way to get at the fundamentals of structure than by successive transformations to strip away the secondary properties," writes James R. Newman in *The World of Mathematics,* in a prelude to a section on symmetry.

> It is a method analogous to that used by the archaeologist who clears away hills to get at cities, digs into houses to uncover ornaments, utensils and potsherds, tunnels into tombs to find sarcophagi, the winding sheets they hold and the mummies within. Thus he reconstructs the features of

an unseen society; and so the mathematician and scientist create a theoretical counterpart of the unseen structure of the phenomenal world.

Whitehead has characterized these efforts in a famous observation: "To see what is general in what is particular and what is permanent in what is transitory is the aim of scientific thought."

Relativity is all about using symmetry to discriminate between the general and the particular—between broad truths that apply to everything and narrow consequences of local conditions. So what is symmetry, exactly? Why does it have such power to reveal truth and create beauty?

A snowflake might seem to be more symmetrical than a sphere. But to a mathematician, a sphere is the most symmetrical geometrical figure around. The reason is, you can transform the sphere in many more ways than you can transform the snowflake and still have it look the same.

A circle is more symmetrical than a snowflake. In the top row, the snowflake to the far right has been rotated through 60 degrees relative to the middle snowflake; it looks the same. The snowflake to the far left has been rotated less (or more) than 60 degrees; it does not look the same. In the bottom row, a circle can be rotated by any amount and you cannot tell the difference. It is perfectly symmetrical (at least as far as rotation about its axis is concerned).

Imagine a sugar cube. If you were to rotate the cube a quarter turn so that first one side, then another side, faced you, you could not tell the difference. You could not tell the bottom from the top, or the top from the sides. But if you rotate the cube an eighth of a turn, so that the corner faces you, it looks different. You can turn a sphere, however, in an infinity of ways and it will always look the same.

Many of the most beautiful patterns created by nature and human nature have a great deal of symmetry—tiling patterns and decorative borders and snowflakes and daisies all take a single simple pattern and rotate it or flip it or turn it upside down.

If the thing stays the same after you transform it, you've found another symmetry. A chemist friend said he once used a game to teach symmetry to his children's classes. He brought in an assortment of shapes, with various kinds of symmetry. One child wore a blindfold, while the second transformed the shape in some way. If the first child couldn't tell the difference in the shape when he took off the blindfold, they'd found another symmetry.

The more dimensions you have, the more symmetries you generally have—that is, three-dimensional objects like spheres have more symmetry than two-dimensional objects like circles. There is more elbowroom in higher dimensions to turn things around and still have them look the same.

Imagine the sugar cube again. Imagine it gets squashed into two dimensions—a flat square on your tabletop. Now there are only four ways the former "cube" can sit and still look the same. The transformation of a sphere is even more dramatic. Imagine a dinner plate—essentially a flattened sphere. If you look at it face on, it looks like a circle. But if you rotate it, edge on, it can look like a line. At various angles in between, it looks like a series of fatter or slimmer ovals. It changes its shape, in other words, depending on the angle it presents to your eye.

The same shape in three dimensions, however, has an infinite number of symmetries. You can rotate a sphere however you like, and it will still be spherical. (By extrapolation, perhaps you can imagine how a three-dimensional sphere that could puff up into four dimensions could contain even more symmetry!)

Curiously, while people are drawn to symmetrical shapes, they find too much symmetry bland, as Ian Stewart and Martin Golubitsky point out in their fascinating book *Fearful Symmetry: Is God a Geometer?* (Their answer, by the way, is yes, She is.) Somehow, we need to crack the perfect symmetry of a boring sphere to see a "pattern." A snowflake seems more symmetrical than a circle because the perfect symmetry has been broken; a snowflake only looks the same if you rotate it about sixty degrees. But a circle looks the same no matter how much it rotates around its center.

Symmetries are not restricted to space. Some things are symmetrical in time. It does not matter whether you look at the sugar cube now or five hours from now, it will look the same. That's a symmetry. It does matter whether you put your socks on before your shoes (or vice versa). That's broken symmetry.

"It is hard to imagine that much progress could have been made in deducing the laws of nature without the existence of certain symmetries," writes physicist David Gross, in his paper "The Role of Symmetry in Fundamental Physics." "The ability to repeat experiments at different places and at different times is based on the invariance of the laws of nature under space-time translations [transformations]."

One of the most mysterious properties of time is that it appears to be symmetrical for single subatomic particles, but unsymmetrical when great congregations of particles get together. An interaction between two particles can look exactly the same moving forward and backward in time; but aggregates of particles—including people—can go only forward, not backward, in time.* You can always tell if a

*See Chapter 6, "Emerging Properties: More Is Different."

movie is running backward because the events it depicts are not symmetrical in time. However, you couldn't tell if a movie showing the interactions of two particles was running backward.

Symmetries also show up in transformations involving context or scale or shape. Mountains and molehills share roughly the same shape, as do swirling stars in galaxies and the swirling cream in coffee. Snail shells and sunflowers repeat the same patterns over and over because their genetic instructions encode a symmetry of proportion: The next row of petals or twist of shell always grows so that it remains in exactly the same proportion to the one that follows and the one that proceeds. "This notion of symmetry . . . ," writes Rothstein, "is close to what we call, in other contexts, harmony."

Still more subtle symmetries come out of a branch of mathematics called topology—a kind of elastic geometry where lines and shapes can be transformed into one another so long as there is no tearing or breaking. Topologically speaking, a coffee mug is essentially the same as a donut because both have a single hole surrounded by a single uninterrupted surface. If a donut were made of putty, you could shape it into a coffee mug without tearing the surface. You couldn't make a sphere into a donut without ripping the surface, however. Donuts and coffee cups share a symmetry that spheres don't. (Mathematicians joke that a topologist is someone who doesn't know the difference between a donut and a coffee cup.)

All of these are symmetries, and all speak of deep connections that lie buried underneath the superficial differences. They are the same kinds of symmetries, Rothstein argues, that create the emotional responses we "feel" in the presence of beauty—in math or music. "What we 'feel' in such moments is the analogy of the part and the whole, object and other object, relation and relation."

You can even search for symmetry when you don't know what you're talking about. That's the beauty of algebra! Take a simple equation like *A* plus *B* equals *B* plus *A*. The statement is symmetrical whether you're talking about apples or oranges, galaxies or frogs. The statement is symmetrical because if *A* and *B* switch places, the outcome will be the same; you can't tell the difference.

A variation of this equation lies at the heart of the Golden Rule, which tells us to do unto others as we would like others to do unto us. It's saying that if the doer and do-ee change places, the outcome ought to be the same. It is also encoded into many kinds of laws, as a measure of fairness. If you cut a cake into two pieces to share with a friend, and the cut divides the cake into two exactly equal pieces, it shouldn't make any difference which piece you choose. The outcomes, in other words, are symmetrical.*

Symmetries can be hard to see on the surface, however, because things do, in fact, appear to be different depending on your point of view. Sitting on Earth, for example, we see the Sun move across the sky each day from east to west. It took an enormous amount of imagination to see that from the Sun's point of view, we move in circles around it. The two views are symmetrical, and the relative motions of the Sun and Earth don't change with your point of view, but what you see certainly looks different depending on where you're standing.

In the same way, it's not obvious that filet mignon and brussels sprouts and people are all essentially carbon and water. It's not obvious that soot and diamond are made of exactly the same ingredient (carbon). You have to look deep inside to see the connection, the sameness, underneath.

Sometimes the internal symmetry gets reflected on the macroscopic scale. The visible symmetry of a snowflake is patterned on the strength and nature of water's hydrogen bonds. It is chemistry made visible—the underlying molecular structure magnified to macroscopic scale. The idea that every carbon atom and water molecule in

*See Chapter 10, "Fair Division: The Wisdom of Solomon."

the universe is exactly like every other is another way to think about symmetry. If you can't tell things apart, then they have to be perfectly symmetrical. In this sense, symmetry is about discernibility. As physicist Philip Morrison points out, "What is symmetrical under one aspect of perception may not be so under another. If I am color-blind, I cannot tell the red-marked side of the boat from the green, the port from starboard. The boat is, for me, perfectly symmetrical." (We sometimes call things that are exactly alike mirror images, but even a perfectly flat mirror image is reversed, left for right—a consequence of the light's angled path from yourself, to the mirror, and back to you.)

Carbon atoms, however, are completely indistinguishable. If a carbon atom on your fingertip spent previous centuries spinning around in the atmosphere or buried at the bottom of the sea in some fossil skeleton, you could not find out its history. No matter what happened to it, it would stay exactly the same—and exactly the same as every other carbon atom in the universe. The universal symmetries of atoms come from even deeper symmetries present in subatomic particles and the glue that holds them together.

Ironically, perhaps the most symmetrical thing of all is nothing. Indeed, physicists sometimes describe "nothing" as a state of perfect symmetry. You can tell you're in "nothing" because it doesn't make any difference which way you turn it or look at it or transform it in any way. To a fish, still water would be roughly analogous to nothing, because it can't tell one direction from another. If the water crystallized into ice, however, then the alignment of the crystals would provide direction; each direction would no longer be the same as any other.

Some physicists think that matter came into being when "stuff" froze out of "nothing" just as a crystal ice cube freezes out of the amorphousness of water. They use mathematics to search for the broken symmetry that turned nothing into us.

And so it goes with Einstein's special relativity. Underneath the relativity of space and time is the absolute nature of the speed of light. If a light beam is racing toward you, and you run to meet it, you will not get there any faster; if you try to run away from it, it will catch you just the same. This is very different from running to get a train, where you could catch up if you could only run fast enough. You can never gain on light.

All the strange elasticity of space and time that emerges from relativity follows from this simple fact. Because speed is distance over time (as in sixty miles per hour), if the speed never changes, then time and space must both be flexible. And the measured speed of light is always 186,000 miles per second, no matter how you are moving.

Another deep (and surprising) symmetry in special relativity is $E = mc^2$, which says that energy and matter are underneath the same stuff and can be transformed into each other at will.* The Sun's nuclear furnace, for example, spews out several ocean liners worth of mass each day in the form of radiant energy. The energy that plants harvest from the Sun turns into (among other things) you.

Gross points out that Einstein's great advance in his 1905 paper on special relativity was "to put symmetry first. . . . This is a profound change in attitude."

Einstein's general theory of relativity covers an even larger canvas. Einstein saw that falling off a building was exactly equivalent to floating in space and being pulled toward Earth by gravity was exactly equivalent to accelerating in a rocket ship. If an astronaut reaches for an apple floating about the galley of the spacecraft and the ship suddenly accelerates, the apple (and the astronaut) will "fall" to the floor. Falling and accelerating are exactly equivalent.

The constancy embedded in symmetry and invariance relates quite directly to what are known as conservation laws—for the good

*Energy equals mass times the speed of light (c) squared.

reason that they spell out which aspects of nature are absolutely conserved, under all conditions. Energy, for example, is a conserved quantity. You can move it around and transform it and even change it into matter (and vice versa) but the total amount never changes.

So is electric charge. In our universe at least, every positively charged particle appears to have a negatively charged counterpart. So the amount of electric charge in the universe always stays the same. You can't create only positive electricity, or only negative, but you can separate them to take advantage of their mutual attraction and power. Thunderstorms separate electric charges in air molecules with a terrifying force that becomes visible as currents crash across the sky, bringing the separated charges back together.

The person who crystallized this critical connection between symmetry, invariance, and conserved quantities in a way that saved general relativity for Einstein was a young German mathematician named Emmy Noether.

Raised to clean house, cook, and go to dances like all the other girls, Noether arrived on the scene just in time to slip through some just opening cracks in the barriers keeping women out of science. (Her father was an eminent mathematician and apparently supportive, which helped. In early reference books on the era, Emmy was listed as the daughter of Max; later books describe Max as the father of Emmy.)

She was not permitted to lecture at the university because she was a woman. The great mathematician David Hilbert tried hard to get her an appointment at the Philosophical Faculty in Göttingen; Einstein wrote a letter in her behalf, but with no result. (Apparently, the gentlemen at Göttingen had not yet figured out—or did not wish to know—that mathematical genius was invariant as to gender. Or as Hilbert apparently argued to his fellows: "I do not see that the sex of the candidate is an argument against her admission as *Privatdozent.* After all, we are a university and not a bathing establishment.")

Despite these obstacles, Noether became a major contributor to mathematics. Almost as soon as Einstein published the general theory, which described gravity as a curvature of four-dimensional space-time, mathematicians set out to explore the properties of this intriguing new territory. The theory was rough and unfamiliar and there were problems. Foremost among them, it appeared that energy wasn't conserved in curved four-dimension space—a fundamental flaw.

Noether solved the problem, using symmetry to prove that energy was conserved in four-dimensional space. But she went far beyond that. (She was not allowed to present the paper containing the theorem herself; it was submitted by mathematician Felix Klein.) Noether's theorem proved that conservation laws are the same as laws of symmetry—a huge breakthrough. Because laws of physics are symmetrical, they do not change over distance or time, in space or on Earth, on large scales or small, today or tomorrow.

As was her habit, Noether reached from the particular to the general, the transitory to the permanent. "Her genius is that she solved it with a depth and generality, not only for general relativity," says UCLA physicist Nina Byers, who's made a study of Noether's contributions to particle physics. "She solved it for all of physics."

Einstein wrote about her work: "It amazed me that one could view these things from such a general standpoint. It would not have done the Old Guard at Göttingen any harm, had they picked up a thing or two from her. She certainly knows what she is doing."

It's fitting that Emmy Noether played such a prominent role because she was the kind of mathematician who sees big, broad, general truths—so the idea of relating symmetry to fundamental laws of nature came naturally. She was not interested in calculation; in fact, she was so far removed from such pedestrian activities that some people called her brand of mathematics "theology."

Hers was a mathematics not of cash registers and recipes, but of truth and beauty. Sharon Bertsch McGrayne writes in *Nobel Prize*

Women in Science that in Noether's greatest papers, "formulas, numbers, physical examples, and computations fade away. It is as if she were describing and comparing the characteristics of buildings—tallness, solidity, usefulness, size—without ever mentioning buildings themselves. Numbers and formulas actually seemed to hinder her understanding of mathematical laws and proofs."

Like Einstein himself, Noether saw the hidden structures that held seemingly dissimilar things together. In honor of her memorial, Einstein wrote a letter to the *New York Times* describing her as a "creative mathematical genius" who discovered methods "of enormous importance."

The most-esteemed scientists of her day mourned her sudden death in anguished terms. It was a sad irony that having escaped the Nazis and having penetrated the barriers against women in academia, she apparently succumbed to an infection that set in suddenly after successful surgery. Still in her early fifties when she died, she was "at the summit of her mathematical creative power," said physicist Hermann Weyl in his 1935 memorial address. He seemed to have captured the general mood when he said that she "was such a paragon of vitality, stood on the earth so firm and healthy with a certain sturdy humor and courage for life, that nobody was prepared for this eventuality."

Noether's ideas about symmetry and natural law embody one of the most concrete examples of the link between truth and beauty—a link that builds on connectedness. "What's beautiful in science is that same thing that's beautiful in Beethoven," said the great physicist Victor Weisskopf. "There's a fog of events and suddenly you see a connection. It expresses a complex of human concerns that goes deeply to you, that connects things that were always in you that were never put together before."

Today Noether's ideas about truth permeate physics. Physicists rely on Noether's theorem, even though many have no idea who Noether was or that she was a woman. "To understand nature, that is, to understand its rules, is equivalent to understanding its symmetries," according to physicist Lawrence Krauss, who wrote *The Physics of Star Trek.* "This is why particle physicists are obsessed with symmetry. At a fundamental level, symmetries not only describe the universe; they determine what is possible, that is, what is physics."*

The search for symmetries has led to, among other things, the discovery that nuclear particles like protons and neutrons contain even more fundamental building blocks called quarks. Quarks were discovered in symmetries before anybody thought to look for them in the debris of particle collisions in high-energy accelerators. Finding the symmetries, physicist Murray Gell-Mann was able to figure out which quantities had to be conserved. The task of putting together all the pieces of that puzzle is still the subject of dozens of international scientific searches.

Indeed, it is quite similar to the discovery of the periodic table of elements—that is, the fact that atoms fall into families in a very ordered way, with clear similarities across generations. Once the family patterns were clear, it was easy enough to figure out which—if any—members were missing, just as it's easier to tell which pieces are missing from a jigsaw puzzle after the puzzle is put together.

Symmetry also led to the discovery of antimatter. This strange stuff—unknown on Earth before the 1930s—made its first appearance as a minus sign in an equation. When physicist P. A. M. Dirac mathematically combined special relativity and quantum mechanics, the marriage produced twin symmetrical solutions—one with a plus sign, the other with a minus—two versions, each the exact mirror image of the other.

Could such a thing as antimatter really exist? Caltech's Carl Anderson found that it could and does—shortly after Dirac made his prediction. Looking at photographic plates of particles streaming in

*This reference is from another of Krauss's popular books, *Fear of Physics.*

from space, he found what looked like an electron curving the wrong way in a magnetic field. It was an antielectron, or positron, the first known to humankind. In 1949 the late Richard Feynman showed that mathematically, an antiparticle is the same as a particle moving backward in time.

Today, antimatter particles like antiprotons and positrons have become routine tools in both particle physics and medicine. PET scans (Positron-Emission Tomography) make use of antimatter to look at the goings-on inside your brain.

In a universe with perfect symmetry, of course, we would have known about antimatter from the get-go, because antimatter would have been everywhere. But then, we wouldn't have been around to notice.

That's because the amount of matter in the universe is also a conserved quantity. When you convert energy into matter, you always get as many antiparticles as particles. And when particles and antiparticles meet, they annihilate in a burst of pure energy.

So the thorny question arises: If the universe came into being by a burst of pure energy, where did all the antimatter go? It must have been there, because the laws of physics are symmetrical. And if there was as much antimatter as matter, then every bit of matter would have joined with a bit of antimatter and annihilated each other into nothingness. That clearly didn't happen since something stuck around to evolve into stars and galaxies and planets and us.

Physicists who try to answer the question of why there is something rather than nothing in the universe study symmetries. Indeed, much of particle physics these days is based on a mathematical notion called group theory—where a group is the collection of all transformations of an object that leave that object invariant. In some corners of physics, there is already good evidence that nature isn't as symmetrical as people thought. Some subatomic particles behave in unsymmetrical ways. The perfect symmetry of the newborn universe must have been broken. A great deal of high-energy physics and cosmology these days is centered on finding out how and why.

Symmetry is also behind much of the excitement about string theory, which physicist Edward Witten of the Institute for Advanced Studies in Princeton likes to call a piece of twenty-first-century physics that dropped into the twentieth century by mistake.

String theory replaces the notion that tiny point particles are the building blocks of nature; instead, the fundamental units are unimaginably smaller vibrating strings. These are not strings of twine, but strings of some unknown fundamental stuff—even more fundamental than space and time. The strings vibrate not only in the usual three dimensions familiar on Earth and the fourth dimension of time, but also in six other dimensions that are rolled up too small to ever see. Their harmonics produce everything that is—which has led some people to call string theory "the theory of everything."

The enormous allure of string theory for physicists grows out of the fact that strings vibrating in ten-dimensional space have enormous amounts of symmetry. They can transform themselves in myriad ways—into gravity and daisies and stars and nerve cells and radioactive atoms—and still stay essentially the same.

Several years ago, string theory begat M theory—known as Magic or Mystery or Mother theory to its fans. M theory is still more symmetrical than string theory, because it incorporates an extra, eleventh dimension. There is even talk of F theory, which would require a twelfth dimension. Today, writes Gross, symmetry serves as the "guiding principle" of forefront physics. "When searching for new and more fundamental laws of nature we should search for new symmetries."

Of course, it's not necessary to go into the exotic world of string theory to see how symmetry is a central theme that comes up over and over in nature. A snail builds its shell according to a symmetry of perfect proportion, sometimes called the golden mean. Vertebrates from field mice to grizzly bears share a central backbone flanked by symmetrical rows of ribs; each eye has a mirror-image partner, as does each five-fingered foot and hand.

But, as Stewart points out, perfect symmetry may be imperceptible. It appears to be broken symmetry that's behind biological patterns as well as cosmological ones. Everything from tiger tails to rose petals result from a breaking of perfect symmetry—just enough breaking so that we see a pattern, but not so much to destroy it completely. "The secret of nature is symmetry," writes Gross, "but much of the texture of the world is due to mechanisms of symmetry breaking."

We ourselves are prime examples of broken symmetry. A baby begins life as a single fertilized egg—almost completely symmetrical. Out of that sameness comes eyes, bones, brain, heart, mind, and music. The question is, if laws of nature are mostly symmetrical, how does all this broken symmetry arise?

Currently, that's one of the juiciest questions in science, and the subject of many excellent books.* But just to give a taste of the problem, consider: Take a bunch of absolutely similar water droplets or nerve cells or stars. How, out of that perfect sameness, do you get snowflakes and thought and galaxies? How do these simple, similar things produce all that pattern and complexity? In the end, it's all a matter of very sophisticated symmetry breaking.

"This is how the Universe spawns complexity from simplicity," writes John Barrow in *The Artful Universe.* "It is why we can talk of finding a Theory of Everything, yet fail to understand a snowflake."

It's also the reason a good many physicists find the whole concept of a Theory of Everything preposterous. After all, physicists could learn everything there is to know about every atom that makes up a

*My personal favorite is *The Collapse of Chaos* by Jack Cohen and Ian Stewart.

living thing and every physical nuance of each musical note, and they still wouldn't understand teenagers or the stirring emotional power of national anthems. Even perfect knowledge of fundamental ingredients doesn't mean you know enough to understand what makes a cake. Symmetry gets lost between cause and effect, says Stewart. Even when the laws of nature are symmetrical, their consequences are not. Rather than explaining the universe, it's fairer to say that a Theory of Everything seeks the ultimate symmetry of the universe, no small accomplishment in itself.

Coming back to human scale, symmetry breaking has wide-reaching consequences for chemistry because many molecules come in two forms—one the exact mirror image of the other. As chemist and poet Roald Hoffmann puts it, they are "the same, and not the same," like right and left hands, right and left shoes. Thalidomide—the tranquilizer that produced such serious deformations in children born to mothers who took the drug—was sold as the two mirror images combined. It seems now that only one version is actually harmful.

Many of these mirror molecules share the same melting points, colors, weights, and so forth. But they may smell different or affect living cells in different ways. One mirror image may be sweet; the other tasteless. One may be a potent painkiller, the other inert.

Biological molecules, in other words, can tell left from right.

As it turns out, most biological molecules have a strong handedness. They twist around like miniature spiral staircases—a symmetry breaking that moves into the third dimension. Right- and left-handed spirals can't be "flipped over" like mirror images to line up exactly, just as no amount of twisting will get a right hand into a left glove.

Yet most DNA as well as proteins spiral in one direction only—to the "right." This right-handed bias is extremely surprising, since nonliving things, such as crystals, come in left- and right-handed forms in roughly equal numbers. (Indeed, exclusively left-handed amino acids on the meteorite from Mars would probably convince the staunchest skeptics that the tube-shaped forms are really fossils of ancient extraterrestrial life.)

Life's innate asymmetry leads to a startling speculation about the origin of life. If life had coiled into existence at many locations at once, then one would expect those early molecules to spiral to the left and right in equal numbers—just as the silicon and oxygen atoms in quartz crystals spiral both ways. But if life originated in a single rare event, then a single molecule would have passed down its right-handed twist as a legacy to all living things to come. In other words, the right-hand twist of so many biological molecules could imply that we have all descended from a single molecule or group of molecules that learned how to replicate itself with ingredients harvested from its environment. We could all share a single ancestor.

As the late physicist Richard Feynman put it, "This fact, that all the molecules in living things have exactly the same kind of [handedness], is probably one of the deepest demonstrations of the uniformity of the ancestry of life, right back to the completely molecular level."

Of course, this is only one of many ideas proposed by the biologists, mathematicians, and physicists who have tried to account for the pervasive right-handed twist to life. Why the asymmetry persists is well understood; after all, right-handed molecules can only react with other right-handed molecules. But how the symmetry got broken in the first place remains a juicy mystery.

Physicist Elsa Feher, who created a wonderful exhibition on symmetry that is now on a three-year tour of science museums in the United States, was wandering one day among her exhibits, just to see how people were reacting. The exhibition is broad and deep, rich with ways for people to play with symmetries in language and music, in space and time, in two dimensions and three, with natural objects and in works of art, with illusions and molecules, human faces and crystals, numbers and tiling patterns.

One exhibit deals with spiral symmetry that rotates into the third dimension around a central point, the symmetry of molecules of life. In the graphics, Feher raises the question of why almost all DNA helixes are right-handed and speculates on the connection between the origin of life and broken symmetry.

On this particular day, Feher stopped next to a man who was playing with the left- and right-handed spirals, reading, engrossed in the sign. He turned to her, thinking she was another visitor, and said: "Imagine that! Did you know that from this, from understanding symmetry in the third dimension, you can make assumptions about the origin of life!"

It got me to thinking about Adam and Eve, the first molecules, and the beginning of life on Earth. It's hardly likely that a geometer god would send us knowledge as a punishment. Far more likely, She sent down an avalanche of interesting broken symmetries—beginning with the difference between woman and man.

SHIFTING FRAMES

What is the real, genuine truth?
To a physicist like me this is an uninteresting
question because it has no physical consequences.
Both viewpoints, curved space-time and flat,
give precisely the same predictions for any measurements.

—*Caltech physicist Kip Thorne*

Anyone who has watched a parade from a parent's shoulders knows the power of changing your point of view. That short scramble six feet up into the third dimension allows whole new vistas to unfurl, and realms that were invisible a moment ago become instantly clear.

All humanity was treated to just such a boost of perspective when the first *Apollo* pictures come back from the Moon, showing Earth as a small watery world alone in endless space. Suddenly our place in the sun was irrevocably altered, even though no new "facts" were presented and nothing had changed but our point of view.

Einstein's relativity is rooted firmly in the idea of shifting frames of reference, as well. Viewing his theories through the lens of reference frames turns invariance on its head. Instead of focusing on what stays the same, it focuses on what looks different—and mostly, how drastically different viewpoints can be valid at the same time.

Indeed, the shifting frames of reference—and the profound idea that reality could encompass both perspectives without either being "wrong"—was the basis of both Einstein's special relativity (the elasticity of space, time, energy, and matter) and general relativity (the curvature of space-time).

This multiplicity of valid viewpoints doesn't imply a return to the "everything is relative" school of relativity. The invariant truths remain, but appearances can be as shifty as shadows.

Shadows, in fact, are as good a way as any to begin exploring shifty reference frames, because they slice reality off at odd angles and send it back to us in tellingly distorted ways. If you walk up the steps to the courthouse on a late summer afternoon, your shadow tags along, but it's not exactly you. For one thing, it's only a two-dimensional slice—perhaps a profile (you would have a nose, say, but no arms), or a front-on view (two arms, but no nose). For another, it's probably stretched out like limp taffy—a consequence of the low angle of the Sun. Most dramatically, it jumps up the stairs at odd angles, folding itself into the contour of the steps, elbowing its way up in a zigzag pattern.

What your shadow reflects, in other words, is a thin slice of you that's been altered by both the light falling on you and the background on which it falls. The three-dimensional object casting the shadow—you—remains invariant through all the transformations.

But your two-dimensional projection morphs into an almost unrecognizable form. (This experiment also offers another good example of how symmetries multiply in higher dimensions. A shadow can be viewed as a two-dimensional slice of a three-dimensional object. The object can remain invariant in three dimensions, but the symmetry is lost in two.)

All these factors—and many more—influence everything we can see or measure. Slices of life can come in any dimension, can be big or small, fleeting or lasting. They can be drawn on backgrounds that are angled, flat, or curved. The part we choose to illuminate, and how, can make all the difference in appearance.

And all reveal a great deal about the limitations and distortions inherent in any single point of view. This is more than mere philosophy. Points of view are quantifiable. If you are moving, things rooted to the ground look different than they do if you are standing still next to them.

These points of view are called reference frames; they put a frame around a bunch of objects or points that are all moving together, say, steadily through space, or conversely, accelerating—like racers coming off the mark. A particular reference frame defines a particular world where things move together, tell time according to the same clocks, are ruled by the same forces. Normally, we take our reference frame for granted; we mistake it for "reality." We rarely stop to think that our everyday reference frame on the surface of good old Earth is whipping us around at thousands of miles a second and propelling us through space at breathtaking speeds.

It's not surprising that changing the way we look at something can have dramatic effects on what we see. In the moving frame of a car, the solid scenery of trees and buildings seems to slip by like a Hollywood set—and we never think twice about it. We know that a person standing by the buildings sees our car rush past at high speed. We accept these mutually contradictory realities as both true—each in its own frame of reference.

What is surprising is how often people deny the power of reference frames to change reality. Take the case of a person sitting in an airplane moving at five hundred miles per hour, tossing a coin. He watches it go up, then down, along the same straight path. To someone sitting on a passing cloud, however, the motion of the coin would appear quite different. The coin would trace a wide, graceful parabola, like water in a fountain—traveling forward as well as up, and forward as well as down.

The person impatiently tossing the coin on the 747 (perhaps waiting for his Scotch) won't perceive the parabola because people traveling on planes don't perceive themselves as moving at five hundred miles per hour—evidence out the window to the contrary. They think they're standing still and use that reference frame as a window through which to view all else.

And therein lies the problem. In order to understand the relationship between what you see and what is going on, you need to add in—or subtract—the influence of your own reference frame. And most people aren't aware that they walk around carrying a frame of reference at all.

Consider the early astronomers who plotted the motions of the planets in the sky. Mars and Venus and the rest of the planetary family all traced baroque curlicues across the sky—like skywriters who'd had too much to drink. These loop-de-loops, or epicycles as they are called, are accurate, serviceable descriptions of the motions of the planets relative to Earth. They served the ancients just fine for making all manner of astronomical forecasts, including solar and lunar eclipses.

The only thing the ancients left out of the planetary orbits was their own frame of reference. They assumed—like the people in the 747—that they were standing still. Shifting the frame of reference to include the motion of Earth and putting the Sun in the center of things eventually transforms the gnarly epicycles into well-behaved ellipses.

If one view is as right as the other, you might well ask, why do people make such a big deal about the enormous intellectual leap made by Copernicus, who rearranged the solar system with the Sun in the middle?* Why do we call the Earth-centered system "backward" and the heliocentric system "progress"? Neither the Sun-centered view nor the Earth-centered view is strictly right or wrong—as long as you know how to translate from one to another, taking your reference frame into account.

The Copernican Sun-centered view is far preferable, however, because the epicycles are so complicated that deep relationships between gravity and the motions of the planets were obscured. Just like the *Apollo* pictures sent back from the Moon, the heliocentric solar system didn't necessarily change any "facts." But its shift in perspective allowed relationships that were muddy to become crystal clear. In particular, it allowed Newton to see relationships between objects and motions that made it possible for him to connect the orbits of planets and moons to the fall of apples on Earth.

Conversely, failure to take your frame of reference into account can lead to illusions (and also jokes). That's because they catch us unawares. For example, the Moon appears larger when it sits low on the horizon than it does overhead because the horizon provides perspective, making the disk appear farther away. Normally, distant objects, like distant sailboats, appear Thumbelina size. So when the Moon appears to be distant, the brain assumes it has to be huge in order to make the same size image on your retina as it does when it's overhead.†

Humorists like Dave Barry lead us into the wrong reference frame to make us laugh. In a recent column, he recounted the amazing events of July 3, 1994, when seven-year-old Jason Toastwanker "fell

*Especially given that Copernicus's new solar system still had a lot of complications. For example, although Copernicus was willing to put the Sun at the center, he insisted that the planetary orbits had to be perfect circles (they are actually ellipses).

†Perceptual psychologists still don't agree on a single explanation for the moon illusion.

off his tricycle, hit his head and was knocked out. When he regained consciousness, he spoke to his parents IN PERFECT GERMAN. This did not surprise them, because they were Germans and this happened in Germany. . . ."

Many times people argue about right and wrong when really what they're contesting is different reference frames. The view that Earth is flat is certainly considered to be wrong these days. And it is wrong, from a global perspective. Everyone saw the *Apollo* photos. And long before that, people were able to deduce the curvature of Earth from the shape of the shadows it made on the Moon and by the way ships sank slowly beneath the horizon.

But the fact of the matter is, Earth is flat for all practical purposes, which is to say, on scales that people normally inhabit. That's because we don't normally cover enough ground in one day to notice the curvature, which is a very long-range affair.

In fact, any curved surface becomes virtually flat if you take a small enough piece of it. Take a soccer ball or even a Ping-Pong ball, cut out a small enough slice, and for all intents and purposes it will be flat. People who thought Earth was flat weren't stupid; they just didn't get out very much. Simply blowing up or shrinking down your reference frame can have enormous consequences.* It's round in one frame, flat in another.

The idea of "reference frame" can apply to many different kinds of concepts that we take for granted to such an extent that we forget they exist. Take the notion of space. We assume space is flat—or perhaps a kind of featureless nothing, like a blank piece of paper or an invisible stage on which the universe "happens." But between Einstein's curved space-time and the foaming quantum mechanical vacuum of subatomic physics, our views of "nothing" have been completely transformed. Space-time has shape and energy; it evolves; it has a past and a future; the subatomic vacuum is alive with the spontaneous genera-

*See Chapter 5, "A Matter of Scale."

tion of so-called virtual particles, matter/antimatter pairs that spring into existence out of nothing, then dissolve back into nothing again.

The background, in other words, plays a role in fashioning the foreground—just as the shape of the shadow changes when it falls on crooked pavement. Everyone who's read a map knows that the shapes and sizes of continents change depending on whether they are represented on a spherical globe or a two-dimensional projection. The frame alters the picture.

The picture certainly seemed simpler before Einstein and quantum mechanics came along and focused our attention on the shape and action of the stage. In those simpler days, geometry took place in featureless space, like the blackboard behind the teacher's desk, where parallel lines wouldn't think to meet and the angles of triangles all added up to 180 degrees. Euclidean geometry subdivided the world into neat triangles, circles, and parallel lines—unchanged for thousands of years.

In fact, James Newman calls the development of non-Euclidean geometry the most exciting mathematical advance of the nineteenth century. For more than two thousand years, the system perfected by Euclid had occupied a position of absolute authority. The rules he had laid down for geometric relations in space were assumed to be as inviolate as the multiplication table. Space obeyed Euclid, mathematicians liked to say, and Euclid obeyed space.

In a way, it's surprising that no one noticed the geometry of curved space sooner. After all, we live on the curved surface of Earth. Any navigator knows that the lines of longitude that are parallel at the equator meet at the poles. And as Thorne points out, two balls dropped on "parallel" paths toward Earth would (if they could) intersect at Earth's center.

For all its familiarity, the idea that the blackboard can bend is exceedingly hard to get used to. There's an old riddle that illustrates this well: Imagine you are standing on Earth, and you walk one mile due south, then one mile due east, then one mile due north, and find

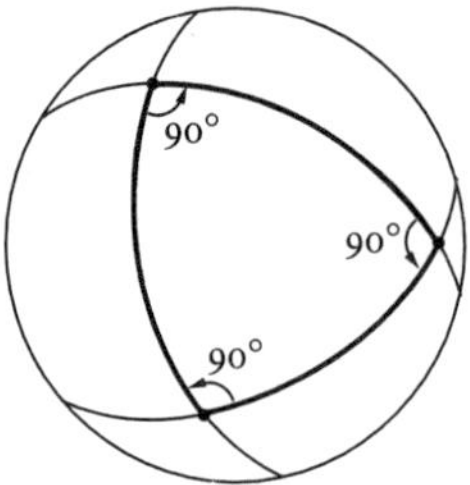

What color are the bears? How three 90 degree angles can form a triangle on the curved surface of Earth.

yourself back at your starting place. What color are the bears? The answer is white, because you are at the North Pole.*

Such is the power of curved space. It can make parallel lines meet and accommodate impossible triangles. And not only on familiar two-dimensional surfaces like Earth, which curve into three-dimensional surfaces like space. According to general relativity, the three dimensions of space mesh with the fourth dimension of time to produce four-dimensional space-time. Imagine it like the threads of a closely woven garment; three dimensions tell you where something is in space; one tells you where it is in time.

The idea that space-time itself curves in the presence of heavy objects—like a water bed when Santa sits on it—is easily accepted these days. It's been tested again and again and passed with flying colors. (More tests are in the works.)

But what about the universe as a whole? Does space-time have an overall shape? If it curves, does it curl up like a ball or bend backward like a saddle?

In Euclid's comfortably flat space, you could assume that the space a million light-years away was just like the space next door, that space 15 billion years ago was the same as space 15 billion years

*There aren't really bears at the North Pole; this just makes the riddle more fun.

from now. The idea that space has shape, however, means it is no longer possible to say with any certainty what is happening everywhere at any time. It could be flat here, but curled up over there. Or flat now, but curled up previously.

Astronomers and cosmologists today use all manner of clever methods to get at the overall curvature of space-time, but it's hard work. There are few fixed reference points in space, and fewer still that come with clearly marked signs, giving useful information such as: "Eight-zillion-watt star. Forty million light-years to Earth."

Even imagining such a thing as the shape of space-time takes ingenuity. Does it go on forever? If not, what's over the edge? Neither answer seems very palatable.

A sneaky way to get a handle on this problem is to shift your frame of reference. Imagine that our four-dimensional space-time is the surface of a balloon. Now, it can have a finite area, but still no edge. Whether this viewpoint is right (or useful) only time (or space) will tell.

Curiously, in Einstein's universe, one's own "time" and "space" were called "proper" time and space, to distinguish them from everyone else's viewpoint. (Needless to say, there are a plethora of social and political situations where we view our own frame of reference as "proper"—and then are surprised to find that everyone else's seems oddly distorted. Many arguments over the need for affirmative action, for example, revolve around the problem of defining a "level" playing field that looks flat to all points of view.)

Our personal perspective includes a host of factors other than the shape of the stage—including time frames, angle of view, the zero point, and a host of others.

Take temperature. The Kelvin scale begins at absolute zero—the coldest imaginable cold—about minus 460 degrees on our familiar Fahrenheit scale. Celsius puts zero at the freezing point of water—32 degrees Fahrenheit. So-called high-temperature superconductors carry electricity without friction at such "warm" temperatures as 130 degrees Kelvin, which is hundreds of degrees below zero Fahrenheit. It's like the old joke: Forty isn't old if you're a tree.

Time frames play powerful tricks on perception as well. Astronomers who think the universe is "young" argue that it's between 8 and 12 billion years old. A "nearby" galaxy is 4 million light-years away. Geologists say the earth beneath our feet flows like a liquid, even though it's solid rock. Whether mountains or continents actually "move" depends entirely on your temporal frame. The elliptical paths of the planets around the Sun had to be discerned by taking many measurements over long periods of time. There is no such thing as an "orbit" over an instant. The point is: things "exist" in some time frames, and disappear in others.

Notions such as hot and cold or old and young, like background and nothing, are meaningless without a proper reference frame. Without context, measurements make no sense. People invent, and reinvent, concepts like zero and nothing and species and organism just as they "invented" the so-called imaginary numbers now essential for dealing with everything from electric circuits to four-dimensional space-time. They aren't a "given" any more than shapeless space or a "second" as a measure of time. (Or as physicist Frank Oppenheimer used to say, frustrated when people would warn him to accept the limitations of the "real world": "It's not the real world; it's a world we made up.")

Comparing human qualities almost always gets muddled by this lack of consensus on measuring instruments and starting points.* Discussing intelligence or worth without universal agreement on the size and shape of measuring sticks is as fruitless as saying the temperature is thirty degrees without specifying whether that's Kelvin, Celsius, or Fahrenheit.

*See Chapter 4, "The Measure of Man, Woman, and Thing."

One of the most intriguing effects of shifting frames is the change from simple to complex, and vice versa. Remarkably simple things flow out of enormous complexity, and vice versa. Many simple similar things (stars, water droplets, neurons) add up to complex ones (galaxies, clouds, minds). As you shift the scale on which you view them, their character changes completely.

Complex things can also acquire elegant simplicity as your breadth of view increases. The family of stars and planets and galaxies, all different in structure and material composition, are all molded into regularly round or roundish shapes. Blood vessels and trees and rivers branch in surprisingly similar ways. The same simple patterns repeat over and over again, growing out of seeming chaos: Jupiter's serene red spot, sitting just below the giant planet's waistline for centuries, is created by unimaginably turbulent storms. Patterns of inheritance, like your father's nose, are recognizable in so many permutations.

The reason these patterns are so robust is that different laws of nature dominate in different reference frames, producing different kinds of behavior.* For example, quantum mechanics only operates at the subatomic realm. Deep inside the table I'm typing on, atomic particles buzz about in probabilistic uncertainties, neither here nor there and everywhere at once; most of the table is empty space. But up here, in my human frame, there is no trace of such goings-on. The table stands solid and sedate.

A building looks simple until you see the wiring and the plumbing. A landscape looks complicated until you look down from a plane and see the repeating patterns of rivers and hills and trees. "Laws of nature are not eternal, abstract truths," writes Stewart. "They are patterns that prevail in some chosen context."

Recently, I was talking to a physicist friend about a problem he was working on: the nature of glasses. Glasses are strange hybrids,

*See Chapter 5, "A Matter of Scale."

solid liquids, if you will, or liquid solids. I mentioned that it seemed to be a complex problem. He responded: "It may just look complex because we don't know the answer. When we know the answer, it may be simple."

The idea of shifting reference frames lends validity to the innate duality of nature, that the opposite of truth, as Frank Oppenheimer used to say, is not heresy. It may be a different kind of truth. Each added view adds insight—so long as the viewer understands the kind of frame he's standing in or the power of the spectacles she's looking through.

Science works, writes Stewart, "precisely because different points of view illuminate different features of the world."

Physicist Viki Weisskopf used to tell a story about two famous physicists walking on the beach. One was lecturing to the other on the mathematical structure of space. The other replied: "Space is blue and birds fly in it."

The point is: both are needed. Or in an oft-repeated truism attributed by various sources to both Christopher Morley and Niels Bohr: "The opposite of a shallow truth is false; the opposite of a deep truth is also true."

Selected Bibliography

Adams, William J. *Get a Grip on Your Math.* Iowa: Kendall/Hunt Publishing Company, 1996.

Axelrod, Robert. *The Evolution of Cooperation.* New York: Basic Books, Inc., 1984.

Barrow, John D. *The Artful Universe.* Oxford: Oxford University Press, 1995.

———. *Pi in the Sky.* Oxford: Clarendon Press, 1992.

Bartlett, Albert A. "Arithmetic, Population and Energy. (Forgotten Fundamentals of the Energy Crisis)." Originally published in the *American Journal of Physics.* No. 9, Sept. 1978.

———. *The Exponential Function* (series). The Physics Teacher, 1976–1996.

Benjamin, Arthur, and Michael Brant Shermer. *Mathemagics.* Los Angeles: Lowell House, 1993.

Brams, Steven J. *Theory of Moves.* Cambridge: Cambridge University Press, 1994.

Brams, Steven J., and Alan D. Taylor. *Fair Division: From Cake Cutting to Dispute Resolution.* Cambridge: Cambridge University Press, 1996.

Byers, Nina. *The Life and Times of Emmy Noether: Contributions of Emmy Noether to Particle Physics.* Proceedings of the International Conference on The History of Original Ideas and Basic Discoveries in Particle Physics: Erice, Italy, 1994.

Carr, Joseph J. *The Art of Science.* San Diego: HighText Publications Inc., 1992.

Casti, John L. *Complexification.* New York: HarperCollins Publishers, 1994.

———. *Five Golden Rules.* New York: John Wiley and Sons, Inc., 1996.

Changeux, Jean-Pierre, and Alain Connes. *Conversations on Mind, Matter, and Mathematics.* Princeton: Princeton University Press, 1989.

Cipra, Barry. *What's Happening in the Mathematical Sciences.* RI: American Mathematical Society, 1995–1996.

Cohen, Jack, and Ian Stewart. *The Collapse of Chaos.* New York: Penguin Books, 1994.

Cole, K. C. *Sympathetic Vibrations: Reflections on Physics as a Way of Life.* New York: William Morrow and Company, Inc., 1985.

Courant, Richard, and Herbert Robbins. *What Is Mathematics?* Oxford: Oxford University Press, 1941.

Coveney, Peter, and Roger Highfield. *Frontiers of Complexity.* New York: Ballantine Books, 1995.

Cronin, Helena. *The Ant and the Peacock.* Cambridge: Cambridge University Press, 1991.

Crutchfield, James P., Farmer, J. Doyne, Packard, Norman H., and Shaw, Robert S. "Chaos." *Scientific American.* December 1986.

Darling, David. *Equations of Eternity.* New York: Hyperion, 1993.

Davis, Philip J., and Reuben Hersh. *The Mathematical Experience.* Boston: Houghton Mifflin Company, 1981.

Devlin, Keith. *All the Math That's Fit to Print.* Washington, D.C.: The Mathematical Association of America, 1994.

———. *Goodbye, Descartes.* New York: John Wiley and Sons, 1997.

———. *Mathematics: The New Golden Age.* London: Penguin Books, 1988.

———. *Mathematics: The Science of Patterns.* New York: Scientific American Library, 1994.

Dewdney, A. K. *200% of Nothing.* New York: John Wiley and Sons, Inc., 1993.

Dick, Auguste. *Emmy Noether (1882–1935),* Birkhauser, 1981.

Eigen, Manfred, and Ruthild Winkler. *Laws of the Game.* Princeton: Princeton University Press, 1981.

Ekeland, Ivar. *The Broken Dice.* Chicago: The University of Chicago Press, 1993.

Emiliani, Cesare. *The Scientific Companion.* Canada: John Wiley and Sons, Inc., 1988.

Farnes, Patricia, and G. Kass-Simon. *Women of Science.* Bloomington: Indiana University Press, 1990.

Feynman, Richard. *The Character of Physical Law.* Cambridge: The MIT Press, 1965.

Freiberger, Paul, and Daniel McNeill. *Fuzzy Logic.* New York: Simon & Schuster, 1993.

Gamow, George. *One Two Three . . . Infinity.* New York: Viking Press, 1974.

Gardner, Martin. *Mathematical Carnival.* New York: Vintage Books, 1965.

———. *The New Ambidextrous Universe.* New York: W. H. Freeman and Company, 1990.

———. *The Night Is Large.* New York: St. Martin's Press, 1996.

Gigerenzer, Gerd. *The Empire of Chance.* New York: Cambridge University Press, 1989.

Gladwell, Malcolm. "The Tipping Point." *The New Yorker.* June 3, 1996.

Glauberman, Naomi, and Russell Jacoby. *The Bell Curve Debate.* Toronto: Times Books, 1995.

Gregory, Richard L. *Mind in Science.* Cambridge: Cambridge University Press, 1981.

Gross, David J. "The Role of Symmetry in Fundamental Physics," *The Proceedings of the National Academy of Sciences,* Vol. 93, No. 25, pp. 14256–14259. Princeton: Princeton University Press, 1996.

Guillen, Michael. *Bridges to Infinity.* Los Angeles: Jeremy P. Tarcher, 1983.

———. *Five Equations That Changed the World.* New York: Hyperion, 1995.

Guinier, Lani. *The Tyranny of the Majority.* New York: Martin Kessler Books, 1994.

Hawking, Stephen, and Roger Penrose. *The Nature of Space and Time.* Princeton: Princeton University Press, 1996.

Herrnstein, Richard J., and Charles Murray. *The Bell Curve.* New York: The Free Press, 1994.

Hoffman, Paul. *Archimedes' Revenge.* New York: Ballantine Books, 1988.

Hoffmann, Roald. *The Same and Not the Same.* New York: Columbia University Press, 1995.

Hofstadter, Douglas R. *Fluid Concepts and Creative Analogies.* New York: Basic Books, 1995.

———. *Metamagical Themas.* New York: Basic Books, Inc., 1985.

Holton, Gerald. "Einstein, History, and Other Passions (Masters of Physics)." Woodbury, N.Y.: American Institute of Physics, 1995.

Horgan, John. *The End of Science.* New York: Helix Books, 1996.

Huff, Darrell. *How to Lie with Statistics.* New York: W. W. Norton and Company, 1954.

Jacobs, Harold R. *Mathematics: A Human Endeavor.* San Francisco: W. H. Freeman and Company, 1970.

Kahneman, Daniel, and Amos Tversky. "The Psychology of Preferences." *Scientific American.* January 1982.

Kasner, Edward, and James Newman. *Mathematics and Imagination.* Redmond, Wash.: Tempus Books, 1989.

Kevles, Daniel J. *The Physicists.* Cambridge: Harvard University Press, 1971.

King, Jerry P. *The Art of Mathematics.* New York: Ballantine Books, 1992.

Kline, Morris. *Mathematics: The Loss of Certainty.* New York: Oxford University Press, 1980.

———. *Mathematics and the Search for Knowledge.* New York: Oxford University Press, 1985.

———. *Mathematics in Western Culture.* New York: Oxford University Press, 1953.

Konner, Melvin. *Why the Reckless Survive.* New York: Viking, 1990.

Krauss, Lawrence M. *Fear of Physics.* New York: Basic Books, 1993.

De Laplace, Pierre-Simon. *A Philosophical Essay on Probabilities.* New York: Dover Publications, 1951.

Lederman, Leon M., and David N. Schramm. *From Quarks to the Cosmos.* New York: Scientific American Library, 1989.

Lewis, H. W. *Technological Risk.* New York: W. W. Norton and Company, 1990.

McGrayne, Sharon Bertsch. *Nobel Prize Women in Science.* New York: Birch Lane Press, 1993.

MacNeal, Edward. *Mathematics: Making Numbers Talk Sense.* New York: Viking, 1994.

Margulis, Lynn, and Mark McMenamin. "Marriage of Convenience." *The Sciences.* September/October 1990.

Margulis, Lynn, and Dorion Sagan. *Microcosmos.* New York: Summit Books, 1986.

Mathematical Sciences Research Institute. *Fermat's Last Theorem: The Theorem and Its Proof, An Exploratorium of Issues and Ideas.* Berkeley: MSRI, 1993.

Morrison, Philip. *Nothing Is Too Wonderful to Be True.* New York: American Institute of Physics, 1995.

Morrison, Philip, and Phylis Morrison. *Powers of Ten.* New York: W. H. Freeman and Company, 1982.

———. *The Ring of Truth.* New York: Random House, Inc., 1987.

Morrow, James D. *Game Theory.* Princeton: Princeton University Press, 1994.

Motz, Lloyd, and Jefferson Hane Weaver. *The Story of Mathematics.* New York: Avon Books, 1993.

National Research Council. *Understanding Risk.* Washington, D.C.: National Academy Press, 1996.

Newman, James R., Ed. *The World of Mathematics.* Redmond, Wash.: Tempus Books, 1956.

Osen, Lynn M. *Women in Mathematics.* Cambridge: The MIT Press, 1974.

Osserman, Robert. *Poetry of the Universe.* New York: Anchor Books, 1995.

Paltiel, A. David, and Aaron A. Stinnet. "Making Health Policy Decisions: Is Human Instinct Rational? Is Rational Choice Human?" *Chance,* Vol. 9, No. 2, 1996.

Paulos, John Allen. *A Mathematician Reads the Morning Newspaper.* New York: Basic Books, 1995.

———. *Innumeracy.* New York: Hill and Wang, 1988.

———. *Beyond Innumeracy.* New York: Vintage Books, 1991.

Perl, Teri. *Women and Numbers.* San Carlos, Calif.: Wide World Publishing/Tetra House, 1993.

Peterson, Ivars. *Islands of Truth: A Mathematical Mystery Cruise.* New York: W. H. Freeman and Company, 1990.

———. *The Mathematical Tourist.* New York: W. H. Freeman and Company, 1988.

Piattelli-Palmarini, *Inevitable Illusions: How Mistakes of Reason Rule Our Minds.* New York: John Wiley and Sons, 1994.

Pickover, Clifford A. *Keys to Infinity.* New York: John Wiley and Sons, Inc., 1995.

Pimm, David. *Speaking Mathematically.* New York: Routledge, 1987.

Poundstone, William. *Prisoner's Dilemma.* New York: Anchor Books, 1992.

Primack, Joel R., and Nancy Ellen Abrams. "'In the Beginning . . .' Quantum Cosmology and the Kabbalah." *Tikkun.* Vol. 10, No. 1. Jan./Feb. 1995.

Root-Bernstein, Robert. "Future Imperfect: Incomplete Models of Nature Guarantees All Predictions Are Unreliable." *Discover* magazine. December 1993.

Rothenberg, Albert. *Symmetry in Art and Science.* Seattle. AAAS Meeting, 1997.

Rothstein, Edward. *Emblems of Mind.* New York: Avon Books, 1995.

Rucker, Rudy. *Infinity and the Mind.* New York: Bantam Books, 1982.

Russell, Bertrand. *Introduction to Mathematical Philosophy.* New York: Dover Publications, Inc., 1993.

Saari, Donald G. *Geometry of Voting.* New York: Springer-Verlag, 1994.

Schattschneider, Doris. *M. C. Escher: Visions of Symmetry.* New York: W. H. Freeman and Company, 1990.

Schimmel, Annemarie. *The Mystery of Numbers.* New York: Oxford University Press, 1993.

Sen, Amartya. "The Economics of Life and Death." *Scientific American.* May 1993.

Smoot, George, and Keay Davidson. *Wrinkles in Time.* New York: Avon Books, 1993.

Steen, Lynn A., Ed. *For All Practical Purposes.* New York: W. H. Freeman and Company, 1988.

Stein, Sherman K. *Strength in Numbers.* New York: John Wiley and Sons, 1996.

Stevens, Peter S. *Patterns in Nature.* Boston: Atlantic Monthly Press, 1974.

Stewart, Ian. *Fearful Symmetry.* London: Penguin Books, 1992.

———. *Game, Set and Math.* London: Penguin Books, 1989.

———. *The Problems of Mathematics.* Oxford: Oxford University Press, 1987.

Stoppard, Tom. *Arcadia.* London: Faber and Faber Limited, 1993.

Tavris, Carol. *The Mismeasure of Woman.* New York: Simon & Schuster, 1992.

Thompson, D'arcy. *On Growth and Form.* New York: Cambridge University Press, 1961.

Thorne, Kip S. *Black Holes and Time Warps: Einstein's Outrageous Legacy.* New York: W. W. Norton and Company, 1994.

Thurston, William P. "On Proof and Progress in Mathematics." *Bulletin of the American Mathematical Society,* April 1994.

Tobias, Sheila. *Overcoming Math Anxiety.* New York: W. W. Norton and Company, Inc., 1978.

Waldrop, Mitchell M. *Complexity.* New York: Simon & Schuster, 1992.

Weinstein, Neil D. "Optimistic Biases about Personal Risks." *Science,* Vol. 246. December 8, 1989.

———. "Why It Won't Happen to Me: Perceptions of Risk Factors and Susceptibility." *Health Psychology,* 1984, 3 (5).

Weyl, Hermann. "Emmy Noether." Memorial address delivered April 26, 1935, at Bryn Mawr. Published in *Scripta Mathematica III,* 3 (1935).

———. *Symmetry.* Princeton: Princeton University Press, 1952.

Wilczek, Frank. "The Cosmic Asymmetry between Matter and Anti-Matter." *Scientific American.* December 1980.

Wooley, Benjamin. *Virtual Worlds.* England: Penguin Books, 1992.

Young, H. Peyton. *Equity in Theory and Practice.* Princeton: Princeton University Press, 1994.

Zaslavsky, Claudia. *Africa Counts.* Boston: Prindle, Weber and Schmidt, Inc., 1973.

Zimmer, Carl. "The Sharebots." *Discover* magazine, September 1995.

Index